Humanity Enhanced

Basic Bioethics

Arthur Caplan, editor

A complete list of the books in the Basic Bioethics series appears at the back of this book.

Humanity Enhanced

Genetic Choice and the Challenge for Liberal Democracies

Russell Blackford

The MIT Press
Cambridge, Massachusetts
London, England

MIT Press books may be purchased at special quantity discounts for business or sales promotional use. For information, please email special_sales@mitpress.mit.edu.

This book was set in Sabon by the MIT Press. Printed and bound in the United States of America.

Library of Congress Cataloging-in-Publication Data

Blackford, Russell, 1954–
Humanity enhanced : genetic choice and the challenge for liberal democracies / Russell Blackford.
    pages   cm. — (Basic bioethics)
Includes bibliographical references and index.
ISBN 978-0-262-02661-1 (hardcover : alk. paper)
1. Genetic engineering—Moral and ethical aspects. 2. Genomics—Moral and ethical aspects. 3. Rational choice theory—Political aspects. 4. Human beings—Psychology. I. Title.
QH442.B53   2013
174.2—dc23
2013015058

10 9 8 7 6 5 4 3 2 1

To Alison Kennedy, who set me on this path.

Contents

Series Foreword

I am pleased to present the fortieth book in the Basic Bioethics series. The series makes innovative works in bioethics available to a broad audience and introduces seminal scholarly manuscripts, state-of-the-art reference works, and textbooks. Topics engaged include the philosophy of medicine, advancing genetics and biotechnology, end-of-life care, health and social policy, and the empirical study of biomedical life. Interdisciplinary work is encouraged.

Arthur Caplan

Basic Bioethics Series Editorial Board
Joseph J. Fins
Rosamond Rhodes
Nadia N. Sawicki
Jan Helge Solbakk

Acknowledgments

Humanity Enhanced is based on my doctoral dissertation, completed at Monash University between 2004 and 2008. I gratefully acknowledge the help of my supervisor, Justin Oakley, throughout that process. His wide knowledge of moral philosophy opened many paths that might otherwise have been closed to me. I also gratefully acknowledge the assistance of Robert Sparrow, my associate supervisor, and Dirk Baltzly, my acting supervisor for much of 2005.

My dissertation examiners, Nicholas Agar and Gregory Pence, responded with a mix of enthusiasm, encouragement, suggestions, and some searching objections. Their comments were invaluable to me as I set about rethinking, revising, and extending the thesis for possible publication.

I am immeasurably indebted to Jenny Blackford for her support, forbearance, questions, criticisms, superior computer expertise, and invaluable proofreading at all stages of this project.

Many other friends, loved ones, and colleagues were kind and supportive throughout all the highs and lows, and I can never thank them enough—in some cases, just for being there. Among those who most encouraged this particular project, or work closely associated with it, were Gregory Benford, John Bigelow, Damien Broderick, George Dvorsky, Ron Gallagher, Alison Goodman, James Hughes, Alison Kennedy, the late Paddy McGuinness, Andy Miah, Chris Moriarty, Simon Smith, and Mark Walker. My shamefaced apologies to anyone else I should have included on this list, but inadvertently overlooked!

The book also has been improved from the efforts of editors associated with the MIT Press—namely, Clay Morgan and Art Caplan—and the comments of anonymous peer reviewers consulted by the Press.

1

Motivation and Overview

Acrimony and anger are all too common in the field of human bioethics. This is evident whether the issues are familiar ones, such as the rights and wrongs of abortion and euthanasia, or those of more recent notoriety. Modern biotechnology in its various guises has transformed much of clinical practice, blurring what once seemed like clear ideas of life and death, and challenging us to reexamine our basic ethical and political principles. Do these principles make sense under new circumstances, and in any event, how should they be applied to sensitive cases at the beginning or end of human life? Change generates uncertainty and even fear—emotions that can foster hostility.

Emerging technologies continue to raise new questions. Some of these technologies involve the manipulation of human genetic material, offering the prospect (or bringing the threat) of genetic choice. If employed effectively, they can determine which people come into existence, and with what genetic potentials. How, then, should the technologies of genetic choice be developed by research scientists, and how should they be used by parents and medical practitioners? Importantly, how should their use be constrained by law? Such issues demand careful, open-minded thought, but this is often in scarce supply.

I propose to focus on what Nicholas Agar calls "enhancement technologies": human reproductive cloning, preimplantation genetic diagnosis (PGD), and genetic manipulation or engineering of human embryos. All these allow choices of genetic traits, though the choice with reproductive cloning is simply to pass on a particular individual's genome (Agar 2004, 7–11, 37, 176). I should also note at the outset that much of my argument could be adapted for other technologies of genetic choice or "selective reproduction" (Wilkinson 2010), such as sperm sorting accompanied by artificial insemination or prenatal genetic testing with the option of selective abortion. Like Agar's (2004, 7–11, 37, 176) *Liberal Eugenics*, however, this book will not examine the bewildering variety of other ways that people might attempt to "better themselves," such as with Prozac or cosmetic surgery. Some of the policy issues with these, too, may be analogous to those concerning the technologies of genetic choice. From time to time, that fact will be relevant in the following pages, but any closer comparison will have to await another occasion.

Dolly and Her Discontents

There has been much acrimony over (human) enhancement technologies, and a large part of it can be traced to February 1997, when a team of researchers at the Roslin Institute in Easter Bush, Midlothian, Scotland, announced it had created a female lamb that was an almost-exact genetic clone of an older sheep (Wilmut et al. 1997). The researchers gave their newborn lamb the whimsical name Dolly—in honor of country music singer Dolly Parton—and she was soon famous as the first mammal to be cloned from another that had already reached adulthood.

Dolly was conceived by the somatic cell nuclear transfer technique (SCNT) in which the nucleus of an ordinary (somatic) cell (in Dolly's case, a mammary cell) is transferred to an egg cell from which the nucleus has been removed. Dolly's nuclear DNA, which contained almost the entirety of her genetic code, was identical to that of the sheep that had provided the somatic cell. In short, scientists were able to grow Dolly from the genetic material in an ordinary cell, with no sperm cells used in the process. This was a dramatic breakthrough in reproductive science, but it had even more startling implications.

The theoretical potential to clone human beings in the same way triggered an immediate storm of debate. Politicians and pundits thundered against the prospect of human cloning. There were loud expressions of repugnance, fears of assaults on human dignity, and outraged denunciations of an impending brave new world. Often the tone was dogmatic; sometimes it was apocalyptic. Across the world, innumerable government reports, essays, articles, and books appeared, considering the moral problems if human cloning ever became practical, and whether it should be prohibited by law.[1]

Within the ongoing debate about SCNT and related technologies, a distinction is frequently made between reproductive cloning, to bring about the birth of a human child, and therapeutic cloning. The former would be directly analogous to the conception and birth of Dolly. By contrast, therapeutic cloning involves the creation of a cloned human embryo for certain other purposes, such as biomedical research, or harvesting cells or tissues to be used in therapy. To complicate matters, some procedures other than SCNT can be referred to as cloning, notably embryo splitting, where an early embryo's cells are split off to produce a number of genetically identical copies. The main focus of debate, though, has been on SCNT.

Even reproductive cloning can be viewed as a relatively conservative use of advancing biomedical technology. Potentially more radical scenarios suggest themselves if we begin manipulating the DNA of our children while they are still embryos, thus altering their genetic potentials, perhaps to increase their cognitive or physical capacities. To say the least, such scenarios provoke discontent, alarm, and ongoing controversy.

Leading sociobiologist Edward O. Wilson (1998, 311) has expressed fears that Homo sapiens might, in the extreme, be superseded by a self-designing, self-directed, bafflingly varied form of life: *Homo proteus*. Margaret Somerville (2007, 144–145) asserts that the genetic manipulation of human embryos would destroy "the essence of their humanness" and indeed "the essence of the humanness of us all." In their coauthored book, *Access to the Genome*, Maxwell Mehlman and Jeffrey Botkin (1998, 129) lament that unequal access to new genetic technologies could "threaten the fundamental principles upon which Western democratic societies are based." Elsewhere, Mehlman (2005, 77–82, 79) describes his nightmare that wealth-based access to radical biotechnology is likely to destroy liberal democratic society. He maintains that it threatens "the destruction of democracy and the enslavement of the species" (Mehlman 2003, 189).

Against all this, it is not my intention to hype the possibilities or suggest, in a rhetorical mode, that dramatic policy responses are needed to assert public control over biomedical technology. On the contrary, a reality check may be useful. As I write, many years have passed since Dolly was born in 1996, but there is still no short-term prospect of producing a human baby by reproductive cloning—not least because there have been significant practical difficulties in cloning primates more generally. We nonetheless are already confronted with controversial uses of genetic technology that demand a political response.

The Idea of a Liberal Democracy

Throughout this book, I'll examine a range of contentions that appear in the current debates about enhancement technologies and genetic choices, but I'll argue for a conclusion dramatically different from those that have dominated the debates. The world's liberal democracies are certainly faced with an urgent problem—perhaps a crisis, though that is an over-used word. But the problem is *not* the emergence of frightening technologies that cry out for a strict regulatory response. On the contrary, *fear* of these technologies has created an atmosphere in which liberal tolerance is under challenge.

But what is a liberal democracy anyway?[2] Ingmar Persson and Julian Savulescu have dealt with this question primarily by an ostensive definition, pointing to "the affluent nations of the European Union, the United States, and members of the British Commonwealth like Canada, Australia, and New Zealand." Thus, they say relatively little about abstract, defining features, such as whatever political principles are shared by these countries, but they do emphasize an ideal that each citizen should have equal rights and liberties that are as extensive as possible, "consistently with all others having the same rights and liberties" (Persson and Savulescu 2011, 494; Persson and Savulescu 2012, 4–5, 42–43).

If we search for more precise identifying features, we might frame broader or narrower definitions of what constitutes a liberal democracy. At one extreme, it might be thought that no society is truly a liberal democracy unless it complies strictly with John Stuart Mill's harm principle whenever it enacts coercive laws. The harm principle is essentially the idea that an individual's liberty should be abridged by exercises of social or political power only in response to actions that cause certain kinds of harm to others (Mill 1974, 68).

For better or worse, however, no country in the world adheres to this notion rigorously. All jurisdictions enact at least some coercive laws that are justified to the public on other grounds. In chapter 2, I will argue that liberal democracies should, in fact, pay considerable regard to the harm principle. For now, we need an understanding of liberal democracies that is less specific and more realistic than strict compliance with any particular principle. Perhaps I can begin by identifying some essential, if rather vague, features.

The "democracy" part of the concept at least involves a system of popular elections, though there are many theories and justifications of democracy that include far more than this. Some theorists see the character of democracy, properly understood, as itself constraining what sorts of arguments should be urged and acted on in political deliberations. For example, Amy Gutmann and Dennis Thompson claim that legitimate political rationales must rely on considerations that are accessible to all citizens. Accordingly, they suggest that the state should not act on (and participants in political deliberation should not urge) a contention "that miscegenation is wrong because God says so in the Bible" (Gutmann and Thompson 1996, 56).

Yet such a claim is controversial. In justifying their position, Gutmann and Thompson note that an argument put forward in political deliberation fails to respect a principle of democratic reciprocity between citizens "if it imposes a requirement on other citizens to adopt one's sectarian way of life as a condition of gaining access to the moral understanding that is essential to judging the validity of one's moral claims." By contrast, "many [other] moral claims" that are made in liberal democratic societies can be assessed and accepted by individuals from a wide range of ways of life (ibid., 57). The point seems to be that laws should not be made on the basis of an esoteric moral system—one that makes sense only from the perspective of something like or analogous to a religious sect—even if this moral system is actually endorsed by the majority of voters.

Many citizens of modern democratic countries might be persuaded to accept such a principle of reciprocity. To me, it appears an elegant and attractive position. Still, it is not obvious that it follows from the nature of democracy itself. Gutmann and Thompson's strategy for ruling out sectarian, "inaccessible," or esoteric moral claims depends on a controversial understanding of reciprocity among citizens, and it is not obvious why this should be accepted by someone who follows and advocates an esoteric code. If their idea of deliberative democracy came into conflict with my esoteric moral code—assuming I had one—I might react by asking what is so great about democracy anyway, or about this deliberative understanding of it.

Perhaps more fruitfully, I submit, we can find independent and strong arguments for the state not to act on certain kinds of moral views. An early version of these arguments can be found in John Locke's *A Letter Concerning Toleration* (1983; first published in 1689). They relate less to the requirements of democratic participation, and more to the fundamental role of the state and its evident record of incompetence and failure when it comes to settling controversies over otherworldly doctrines, including moral claims that are closely associated or entangled with them (Blackford 2012, esp. chapters 3 and 5). Once this point is acknowledged, it creates an intellectual pressure for the state to withdraw from arbitrating or enforcing the "true," comprehensive, all-things-considered morality (ibid., 68–73).

Like that of Gutmann and Thompson, this approach is controversial; it will certainly not convince everyone. Yet even somebody who favors an esoteric, perhaps otherworldly, moral code might be impressed by the havoc that has often resulted in the past when rival sects have fought for political power. One response is to step back, seeking a degree of detachment from the sectarian strife. This can lead us to question whether the state should have a role in favoring particular sects over others, or whether it might not do better to sustain a political framework in which people with many differing views of the world can cooperate and live in harmony—or at least with an attitude of mutual toleration. Such an attitude of pluralism and toleration can be bolstered by a general sentiment that no one worldview or conception of the good should be imposed by political power, and that the state should be as deferential as possible to citizens' choices about matters of great importance to them in their own lives.

In a recent defense of liberalism, Paul Starr offers a rich description of the liberal ideal. Liberalism as Starr understands it has allowed people with different religious and moral commitments not just to live side by side but also to flourish together. The state will not require everyone to worship in the same way, follow the same way of life, or profess an official ideology, but it expects citizens to show reasonableness and openness to ideas. It is not neutral about such values as disease and health, sloth and effort, deceit and integrity, cowardice and courage. There are, Starr (2007, 176–177) suggests, excellences that it must promote in order to survive.

As Starr views the situation, the state's agencies and officials regard each person as worthy of being treated equally; value public health and the environment; cultivate character and intelligence; and attempt to educate citizens in integrity, perseverance, empathy, and responsibility. Within that broad framework, the state adopts an attitude of tolerance: it allows for individuals' free development, and accepts a wide diversity of cultural and moral practices. This liberal tolerance may be explained on the basis of individual rights, or equal respect for persons of different faiths and values, but it also has the potential to promote stable cooperation and foster state power (ibid., 85–116).

However we justify it, the "liberal" part of liberal democracy involves widely understood limits to what governments may legitimately do—even with majority support. These may or may not have the status of constitutional law, but in any event they are expressed in such ideas as freedom of religion, freedom of speech, sexual privacy, and a proscription of arbitrary punishments. In societies that have embraced liberal ideas, approval is frequently expressed for Mill's (1974, 120) stance that "experiments in living" should be welcomed. It is assumed that many ways of life are, at the very least, tolerable. While political power will be used for a variety of ends, most ways of life are accommodated to the extent that social peace allows.

Hence Max Charlesworth writes, "In a liberal society personal autonomy, the right to choose one's own way of life for oneself, is the supreme value." He adds that this includes what he calls ethical pluralism: members of the society are free to hold a wide range of moral, religious, and nonreligious positions, with no core values or public morality that it is the law's business to enforce (Charlesworth 1993, 1).

Bioethics and the Challenge to Liberal Tolerance

Illiberal Responses to Genetic Choice

If we take this picture of modern liberal democracies at all seriously, it might suggest that they would adopt a forbearing, even welcoming, attitude to new or foreseeable forms of biomedical technology. We might expect Millian ideas to play a large part in the formulation of policy—and these would apply to deliberations about the full range of technological innovations, even though Mill himself could hardly have predicted them. For example, reproductive cloning might be welcomed within a liberal democracy as an acceptable personal choice—particularly as a legitimate response to some kinds of severe male infertility and to meet the preferences of some lesbian couples (Blackford 2005, 11–12).

Admittedly, (human) reproductive cloning has never been carried out, and it would involve a high risk of congenital deformities. Reproductive cloning is not a mature and safe technology, and it is not likely to be in the near future. That is a good reason not to attempt it under the current circumstances. Similar arguments, raising safety concerns, could also apply to other technological interventions, such as the genetic engineering of human embryos. Arguments about safety or even the welfare of growing children, do not, however, explain the character of the international debate that we've seen since 1997, or the real-world policy responses of supposedly liberal societies. The debate appears to be motivated in large part by a wish to impose certain moral or quasi-religious ideals as social norms, and by a fear of the potentially strange directions that societies might take as biomedical technology develops.

After Dolly's birth was announced, numerous jurisdictions in the liberal democracies of the West moved to ban reproductive cloning and many other manipulations of human embryos. Even therapeutic cloning is banned in many Western jurisdictions. Italy, for instance, banned all forms of human and animal cloning in 1997. In this case, the ban on animal cloning was eventually dropped, but more comprehensive and stringent regulation of biomedical interventions involving human embryos was introduced in 2004. The new legal regime survived a referendum in the following year, although in 2009, the Italian Constitutional Court struck down some provisions involving assisted reproduction (Metzler 2011). Italy continues to maintain a highly illiberal regulatory regime in respect to all forms of assisted reproduction and embryo experimentation.

In the United Kingdom, the Human Reproductive Cloning Act 2001 prohibits reproductive cloning. The United Kingdom has created a statutory body, the Human Fertilisation and Embryology Authority (HFEA), with a sufficient range of powers to oversee all actions by relevant clinics and laboratories. While UK law provides for stem cell research and therapeutic cloning, under license, the HFEA has consistently opposed the use of PGD for sex selection (Harris 2007, 149–150; Gavaghan 2007, 128; Wilkinson 2010, 213–217).

At the level of European law, Article 3 of the Charter of Fundamental Rights of the European Union forbids what it calls "eugenic practices" and especially "those aiming at the selection of persons." It also contains a specific prohibition of human reproductive cloning. How exactly these can be rights is somewhat mysterious—who holds these rights, and against whom are they held? If anything, such provisions seem aimed at *removing* the right of parents to use particular technologies. In any event, legal policy at the level of the European Union leans heavily against any manipulation or instrumental use of human embryos.

In my own country, Australia, one outcome at the federal level has been the enactment of the Prohibition of Human Cloning for Reproduction Act 2002.[3] In its current form, as amended, this act specifies a raft of criminal offenses with maximum imprisonment terms of fifteen years—surely appropriate only for major crimes. These offenses include any deliberate alteration of a human cell that would be inheritable, and intended as such (section 15), and any action that involves placing what the act calls "a human embryo clone" in the body of a human being or another animal (section 9). The legislation proscribes germ line manipulation of embryos for any reason, along with actions needed for reproductive cloning.

This federal statute operates in conjunction with numerous other restrictions under state legislation along with guidelines issued by the National Health and Medical Research Council (NHMRC). In particular, the use of PGD for embryonic sex selection, a technology that is already available, is proscribed by NHMRC guidelines and prohibited by some state provisions such as Victoria's Infertility Treatment Act 1995. The latter provides a maximum two-year penalty for any attempt to predetermine the sex of a child by technological means, except where necessary for medical reasons (section 50).

In the United States, there has been an extraordinary flurry of legislative activity at both the federal and state level. As a result, a broad range of US states have enacted prohibitions of human cloning and regulation of embryo research of various kinds. The relevant statutes invariably prohibit reproductive cloning, and often prohibit or restrict therapeutic cloning or embryo research in general.

Initiatives have also been taken in many other Western liberal democracies as well as industrially advanced Asian countries such as Singapore and South Korea to prohibit reproductive cloning, and impose other limitations on genetic and reproductive technologies. In short, the broad direction of public policy in the world's economically developed countries has been *against* liberal tolerance of existing technologies such as PGD or sperm sorting for sex selection, the possibility of reproductive cloning, and certainly the more outré and speculative possibilities of genetic engineering to enhance human potential.

Illiberal and "Liberal" Arguments

Nor has this tendency been without high-profile support from academic thinkers and public intellectuals. On the contrary, restrictions on innovations in biomedical technology have been pressed energetically by a wide range of critics. Some of these are plainly out of sympathy with the whole idea of liberal tolerance; others attempt to head off specific innovations in what they evidently regard as special circumstances that might arise from disruptive technological change. The line between these categories of critics may not be sharp, but there is no doubting that some rely on arguments that are clearly illiberal.

What makes them so is the combination of two things. First, there is an appeal to personal moral views that may (or may not) be popular or psychologically attractive, but have no claim to privileged authority in a society that accepts reasonable ethical pluralism. Second, the advocates of these arguments do not seek merely to persuade others to act voluntarily. If that were the aim, they could supplement their abstract claims about morals and moral virtue in other ways: they could lead by personal practice and example; associate with like-minded people; conduct awareness campaigns; or even refuse to be employed by some organizations, such as universities involved in certain kinds of biomedical research. Modern liberal democracies, which value freedom of speech and association, provide ample opportunity for all these actions. But these critics wish to go beyond persuasion; they advocate the employment of political coercion to control the behavior of those who might disagree.

Somerville is one of many who approach bioethical questions with such an illiberal spirit. The direction of her work is to argue against a wide range of innovations on the basis that they fail to respect life in the correct way or that they are contrary to nature.[4] Such reasoning leads her to insist on legal prohibitions, even when our human sympathies might incline us the other way and no obvious harm is involved. Consider her approach to the idea of sex selection for so-called family balancing. Somerville insists that the law should say "no" to a couple who have three boys and want to have a girl, even though she thinks that saying "no" requires more "courage" than saying "yes." The illiberality is more apparent when she adds that we must act because the value embodied in the law will become the societal value (Somerville 2007, 137–138).

To say the least, this is confused. For starters, the only reason why it might be thought to require more courage to say "no" is that it requires denying something to a couple whose deepest wishes will be set back by the refusal. It is not courageous, to put it bluntly, in the sense of placing the naysayers at risk for a good cause; it is only so in that there may be a kind of courage in insisting on one's convictions and imposing them on others, even when doing so causes suffering. That, I submit, is not the sort of courage that should be reflected in our laws.

Somerville's statement is also confused because it falsely assumes that merely *allowing* a practice—that is, declining to prohibit it—entails its endorsement by society as a whole. Liberal democracies should never accept that proposition. They tolerate many practices that various individuals—sometimes enough of them to form an electoral majority—disapprove of for religious or moral reasons. Hence, a liberal democracy's mere failure to prohibit a practice does not entail that the practice, along with whatever suspect value or values might underlie it, is officially endorsed. For example, Charlesworth is on strong ground when he observes that the decriminalization of suicide does not entail that the state morally endorses it. All that follows is that the state categorizes suicide as a matter of personal morality that should not be controlled by law (Charlesworth 1993, 39).

Indeed, even this concedes too much to Somerville, since the state sometimes has good reasons, all things considered, to allow practices that fall outside the strict zone of liberal tolerance. There might, for instance, be problems of enforcement that would lead to evils outweighing the protective benefits of prohibition. Alternatively, it might be unwise and inhumane to stigmatize essentially well-meaning people as criminals. In other cases, legislation might simply be unnecessary or premature. Liberal tolerance can often provide a *good* reason to permit actions that society doesn't endorse, but this does not make it the *only* such reason. The notion that everything not forbidden by the laws of a liberal democracy is thereby socially endorsed is thus doubly wrongheaded and misleading.

It is no doubt possible that a practice will eventually become popular, rather than marginal, if it is not actually prohibited. If that happens, the practice might in a sense *then* be seen to have gained social endorsement. But this is never an inevitable development, and such endorsement as the practice gains is the result of its appeal to many people, not its mere legality. The opponents of new practices have every opportunity to persuade others to refrain from adopting them. Where a practice is new on the scene, has only minority support, and may be contrary to traditional norms, its opponents clearly enjoy great competitive advantages over its adopters and advocates (Feinberg 1988, 66).

More troubling to those of us who take a broadly welcoming attitude to emerging biomedical technologies is the opposition from thinkers whose arguments display a more plausible liberal pedigree than those found in Somerville's writings. Such contentions cannot be so easily dismissed as incompatible with the values that underpin the operation of liberal democracies. Indeed, some of the critics have insisted that certain practices, particularly those involving genetic engineering, are actually a threat to liberal ideas of equality and autonomy. As I will explain in chapter 4, Jürgen Habermas is a key example. Others have alleged dangers to democracy or distributive justice.

Throughout this volume, I propose to concentrate on arguments that appear to carry liberal credentials. These arguments rely on such concepts as welfare and harm, autonomy, justice, and equality—concepts that are familiar in the tradition of liberal thought. Most typically, they emphasize (alleged) harms of some kind—or at least threats to worldly interests of some form—or concentrate on a supposed inconsistency between the impugned technologies and the continued success of societies grounded in liberalism, or of liberalism itself.

The Plan

The chapters that follow consider enhancement technologies in a restricted sense, since the term could apply to a wide range of possible technological advances. In fact, the many possibilities, with specific issues relating to each of them, could sustain a lifetime's research in moral and legal philosophy. Rather than attempt such an encyclopedic project, I will confine the discussion to reproductive cloning, PGD, and genetic engineering, all in the context of genetic choices involving human children. References to (human) genetic engineering are to attempts to modify the DNA of human embryos so as to alter their genetic potentials.

At this point, some readers may suspect that I am therefore ignoring a philosophical distinction between the terms enhancement, in a more technical sense, and therapy. They may even think that this distinction is an important one for morality and/or regulatory policy.

I indeed will distinguish wherever necessary between genetic modifications aimed at eliminating the potential for sickness and those aimed at boosting, or "enhancing," valued characteristics such as intelligence, musical talent, or beauty. Note, however, that the distinction is less than straightforward. For example, boosting intelligence might intuitively seem to be therapeutic for someone whose cognitive abilities or potentials fall below a certain threshold, and otherwise, I suppose, we might regard it as enhancement. So should we think of intelligence boosts as enhancement only when the individual's existing or potential intelligence is already within, or above, something that we regard as the "normal" range? If it falls below that range, does it matter whether or not there is some identifiable organic cause? In any event, even if we can draw a line between therapeutic and enhancing interventions aimed at boosting intelligence, why should this line have any particular significance when we make moral judgments or consider regulatory issues?

I doubt that the much-debated therapy/enhancement boundary is what is truly crucial in current bioethical debates, or that there is a clear, unequivocal boundary to be found. Yet I am conscious that this is a controversial viewpoint. Accordingly, I have provided an appendix in which I discuss the problem at greater length.

Meanwhile, the three technologies bracketed together by Agar have been highly contentious, and much of the discussion that surrounds them contains lessons for the consideration of other innovations that might transform human bodies or psyches, or alter human capacities. Serious consideration of genetic engineering, in particular, inevitably raises questions about the specific uses to which it might be put, whether to prevent disease, or increase cognitive or physical abilities (or boost characteristics that may not actually be abilities, such as physical beauty, however it is understood). For all we know, genetic engineering may eventually enable us to alter psychological propensities or extend the span of human life. Inevitably, what follows touches on such cases; in fact, the possible uses for genetic engineering are so varied, at least in principle, that this technology will dominate much of my exploration.

In chapter 2, I consider the bases for legal prohibitions in liberal theory, the nature of "harm," and whether, if we follow such thinkers as Mill and Joel Feinberg, any legally cognizable harm can be done by the technologies under consideration. This will require some rather technical analysis—perhaps the most difficult material in the book—of the so-called nonidentity problem. In essence, how should we view acts that supposedly harm someone *who would not even exist* but for the act in question? In chapter 3, I focus more specifically on what harm might be done by attempts to modify the DNA of embryos so as to increase potentials for such characteristics as intelligence, longevity, and strength. Most importantly, I argue that some enhancements of human capacities would be genuinely beneficial. In these chapters, I also draw the controversial conclusion, in disagreement with Agar, that we can imagine social circumstances in which it would be justified to exert legal or at least moral pressure on parents to undertake some genetic interventions.

Chapter 4 responds to claims, most prominently associated with Habermas, that the genetic engineering of human embryos poses a threat to the autonomy of the resulting children, and that this, in turn, is a threat to liberal social arrangements. While an investigation of these assertions does leave some residue of concern, it is not sufficient to justify the broad legislative prohibitions that are advocated by Habermas and others.

Chapter 5 deals with the popular idea of an inviolable natural order. Although arguments against violating nature are not usually couched in terms that should be acceptable as a basis for regulatory policy in a liberal society, Stephen Holland has made an attempt to rehabilitate them. As reformulated, the argument relates to harms that are supposedly experienced if certain background conditions to human choice are threatened, confronting some individuals with the prospect of a loss of experienced meaning in their lives. I conclude, however, that the contention lacks sufficient force to overturn ordinary liberal assumptions.

In chapter 6, I consider less direct and tangible harms that might result from human enhancement. As with chapter 4, detailed analysis of the prospects leaves a residue of concern, but not sufficient to justify any sweeping prohibitions in the current circumstances (or to alter the impression that the development of human technologies may, on balance, be beneficial). While there may be some threats to social stability, they are easy to exaggerate.

In chapter 7, I turn to one of the thorniest issues in current debates about human enhancement: the claims of distributive justice. In particular, these claims have underpinned some calls to ban attempts to boost the genetic potentials of children. One difficulty here is that distributive justice arguments are not usually employed in such a strong way in other areas of policy debate, keeping in mind the many other advantages that wealthy parents routinely provide for their children. Why should benefits from genetic engineering be an exception? I hold that there indeed may be something special about genetic engineering, but that in any realistic situations, the argument has only limited force as an appropriate influence on policy formulation.

Chapter 8 considers the implications of the entire discussion for regulatory policy, bearing in mind the values that underpin liberal ideals. Here, I take due account of concerns left over from examinations of more specific assertions and issues. I argue against draconian uses of state power to suppress human enhancement, while contemplating the possibility that some specific and limited regulation might be required in the future. In the current circumstances, though, there is more urgency in simply reaffirming liberal values. In a sense, the continuing debate that followed Dolly's cloning has wasted a golden opportunity.

As previously mentioned, I will conclude with an appendix focused on the much-debated therapy/enhancement boundary, maintaining that it fares badly in both clarity and moral importance.

Conclusion

A remarkable feature of the policy debate that was sparked by the cloning of Dolly, and that now extends to a wide range of emerging technologies, is that many participants have evidently ignored or even rejected liberal approaches to public policy, especially the idea that experiments in living are to be welcomed. All too often, the character of the debate is in conflict with the idea of liberal tolerance, which holds that it is not the proper business of the state to make us virtuous or establish a common morality. Put bluntly, liberal tolerance is under challenge.

But the challenge does not come solely from thinkers who reject it as an important value, and *this* phenomenon in particular demands investigation. Are there features of the emerging technologies of genetic choice that should cause the governments of liberal democracies to weaken, or at least redefine or circumscribe, their commitment to liberal tolerance? My argument is developed in the following chapters. I conclude that these technologies will not be without problems, but they will also bring benefits. Liberal democracies can find a place for them.

2

Human Enhancement and the Harm Principle

One might initially think that liberal democracies would tolerate technologies that offer genetic choices. To be sure, though, the issues are not straightforward. In some cases (e.g., those involving PGD), it is difficult to make sense of the claim that anyone has been either harmed *or* helped by the use of enhancement technologies. If not for a decision to use the technology, the people concerned would not have come into existence in the first place! If we hadn't taken the "harmful" or "helpful" action, those people would not have existed at all.

Such issues require considerable teasing out. I will contend in this chapter that the employment of enhancement technologies should, indeed, be tolerated, at least in large classes of plausible cases. Further, I'll conclude that one element of our ordinary thinking, discernible in all human societies, is a kind of (mild) perfectionism about procreation and parenting. Without adopting any stronger principle, such as the principle of procreative beneficence that Savulescu (2001, 414–418) has argued for, we should at least impose some moderate requirements that relate to the welfare of children.

Savulescu's principle would require couples (or individuals where relevant) to select the child, of those possible to them, with the best or equal-best life prospects, based on the relevant available information, including genetic information obtained by PGD. As will become clear, I think that is going too far. Consider, by analogy, environmental interventions: we don't normally require parents to provide their children with the best interventions possible, such as the finest possible education, but instead set a reasonable minimum. Such an approach makes sense not only for how we treat existing children but also for decisions about which children we should bring into the world.

The Harm Principle

The Scope

Ideally, and to a large extent in practice, liberal democracies tolerate a wide range of religious and moral viewpoints, without attempting to impose the beliefs of one component of the population (even a majority) on others. The role of the state in a liberal democracy is seen as strictly secular: the state's officials and agencies exist to protect and promote worldly interests, rather than to promote spiritual goods, impose otherworldly philosophies, or enforce esoteric canons of conduct. Once that is accepted, there is a pressure for the state to defer to the choices of its citizens wherever possible, including their diverse values and visions of the good. The state increasingly needs to justify its coercive actions by showing how it protects citizens from dangers to their worldly interests, such as those relating to their lives, liberty, and property.

As this logic plays out, Mill's famous harm principle has an obvious attraction. It sums up the idea of a state with limited responsibilities and powers. In *On Liberty*, Mill (1974, 68–69) famously asserts that only the prevention of harm to others can justify the exercise of power over an individual in a civilized community. As elaborated by Mill, this rules out the use of social power (including but not limited to legal sanctions) to intrude on actions that directly affect only the individual concerned and any consenting parties. In Mill's (ibid., 71) view, moreover, it is not sufficient that others might be affected and perhaps harmed through an indirect process.

Before we continue, it is well to remember that criminal law is not the only means by which the state could react with hostility to a practice, attempting to suppress it. In the current social circumstances prevailing in Western societies, criminal law uses punishments that can include the infliction of a range of harms, such as the loss of liberty or property, while also expressing public resentment, indignation, reprobation, and disapproval (Feinberg 1970, 98). But much the same infliction of harm and officially sanctioned stigma could be accomplished by means that do not involve a criminal justice system as we know it. In principle, the state could select many hostile and repressive means to achieve its aims. These include propaganda campaigns that stigmatize certain categories of people and officially tolerated discrimination against people of whom it disapproves.

We therefore should ask for justification whenever the state attempts to suppress behavior or new technology, whether or not its means are technically those of criminal law. Though much of what follows is written, for convenience, as if the issue were confined to criminal prohibitions, the deeper point is that liberal democracies require good justification before they call on the state's power to suppress any form of conduct. Such justification must amount to more than the current government's (or electorate's) wish to support and impose a particular conception of the good.

An Elaboration

The harm principle is also usually confined to the prevention of harms that are considered wrongful or illegitimate (e.g., Mill 1974, 163–164; Feinberg 1984, 35–36).[1] The notions of legitimacy and illegitimacy in play here are difficult to define clearly. Mill and his followers argue, or perhaps assume, however, that a degree of competition or conflict can be acceptable, even if there are adverse outcomes for individuals who lose out. Mill's discussion excludes the harm suffered from the lack of success in a competitive examination or overcrowded profession, unless the means involved have included fraud, treachery, or force.

In my *Freedom of Religion and the Secular State*, I touch on the problems raised by Mill's requirement that the harm be direct, since, as I point out, "it must be recognized that directness is a matter of degree." There can be gray areas, so perhaps we should "think in terms of an urgent risk, or a clear and present danger," instead of making "an inflexible distinction between direct and indirect harms" (Blackford 2012, 74). Perhaps what really matter are such things as the urgency of a situation, likelihood of harm eventuating, or futility of attempts to avert the risk by relatively noncoercive means, such as education or persuasion.

Feinberg (1984, 1985, 1986, 1988) has explored the ramifications of the harm principle in detail and depth in his four-volume magnum opus, *The Moral Limits of the Criminal Law*. His views on the important issue of prenatal forms of harm are further refined in his essay "Wrongful Life and the Counterfactual Element of Harming," originally published in 1986 (Feinberg 1992, 3–36). I do not propose to summarize Feinberg's views in their entirety, but much of his account would be captured by an insistence that any harm be significant, direct or urgent, inflicted on others (rather than self-inflicted), and wrongful. They must also be harms of a secular kind—that is, to worldly things such as life, health, liberty, and property, as opposed to spiritual salvation, holiness or purity from sin, or the gratification of a deity.

Though there is room for a discussion of the fine points, the prevention of harm, understood in approximately this way, has become the least controversial of the possible uses of criminal law (Buchanan et al. 2000, 205). Yet some theorists broadly in the tradition of Mill do recognize a limited role for other principles. For example, Feinberg (1985, 26) accepts offense to others as a legitimate basis for criminalization in some limited circumstances. These situations are likely to involve extreme and widespread repugnance that could not have been anticipated or avoided by those who are susceptible. There might also be countervailing interests for those who are giving the offense, and sometimes these are significant enough to prevail, even if many people find certain public behavior offensive.

Some liberal theorists, such as H.L.A. Hart (1963, 30–34) and Joseph Raz (1986, 22–23), also propose a limited role for the state in protecting citizens against self-inflicted harms. We should, I submit, be wary of such paternalism. It seems more acceptable when those who are protected from themselves do not seem competent to decide for themselves, as may often appear to be the case with children, young teenagers, and elderly people suffering from age-related forms of cognitive decline. Likewise, paternalism may be acceptable where the law imposes a relatively trivial burden, more as a reminder than a punishment, or where, as with the regulation of prescription drugs, exceptional technical competence is needed to make sound decisions. Generally, though, as I observe in *Freedom of Religion and the Secular State*, "competent adults and mature minors have good reason to trust their own decisions about how they live their own lives, and to distrust the efforts of the state to make decisions on their behalf" (Blackford 2012, 76).

This brief look at the content of the harm principle is not definitive, but we can draw some reasonable conclusions about when political coercion is most readily justified:

If coercion can be justified in respect of some indirect harms, it is to the extent that they resemble more direct ones in the need for an urgent response; if it can be justified in respect of some kinds of "mere offense," it is to the extent that the impact of offense merges with that of unequivocal harm; if it can be justified in respect of some self-inflicted harms, it is largely to the extent that we have good reason not to trust our own judgment in areas requiring sophisticated technical competence. (ibid.)

While all these issues are controversial, the harm principle is central for any justification of political coercion in modern liberal democracies. Other rationales will at least have to rely on this-worldly considerations, and should avoid dependence on esoteric moral doctrines.

The possible exceptions to the harm principle examined up to this point seem to have little application to enhancement technologies, although I will discuss indirect harms in later chapters. By and large, enhancement technologies are not opposed by their critics on the basis of up-close offense, such as Feinberg explores, or paternalistic grounds. The next set of issues is of central importance, however: What harm can we do if we bring people into the world using such means as PGD following in vitro fertilization (IVF), or if we were able to use human reproductive cloning? These people would not even *exist* except for our decision to use the technology. Can we harm people whose very existence is contingent on the acts that supposedly harm them?

This is known as the nonidentity problem.

Nonidentity Issues and Detrimental Choices

The Problem

At least initially, some problems about harms to future people seem straightforward. Take the following classes of cases, studied by Feinberg (1992, 11–12). In one class, the harm occurs between conception and birth, as when a negligent motorist runs over a pregnant woman, injuring the fetus, with the ultimate effect that a child is born with a disability. A variation on this theme is that the harmful act may take place prior to conception, as when a socially irresponsible manufacturer produces and distributes medicine prior to conception, but the drug is used *after* conception, with the same result that the child is born disabled.

In the latter instance, it might be suggested that, in reality, the harmful act takes place after conception—that is, if the harmful act is thought to happen when the drug is taken. But more than one act in a chain of causation can contribute to harm, and what, in this case, should we say about the moral or legal responsibility of *the manufacturer*? If we assume that it has been culpably negligent in some way during the process of manufacturing or distributing the drug, and if this negligence took place before the conception of the particular child, should the child be able to respond by suing the manufacturer for the disability that she suffers? If not, why not? After all, the disability would not have eventuated without the manufacturer's actions, including (let us suppose) those that we regard as socially irresponsible.

Perhaps such cases raise philosophical issues; for example, we might get into problems about personal identity if the characteristics of the child are drastically changed from what they would otherwise have been. Still, I feel some inclination to say that the child has been harmed. Given the facts provided, it makes intuitive sense—I suggest—that the child would have been better off if not for the manufacturer's negligent act. That is, she would still have been born, but without the disability.

On the other hand, *would she really*? Who gets born—exactly which sperm cells and ova join to create zygotes that end up developing as embryos, and coming to term as babies—may be *very* sensitive to prior circumstances. Is it really true that *this* exact same child would have been born in a different world where different decisions were made by a particular manufacturer, with all sorts of unpredictable flow-on effects? Public policy might favor setting such questions aside, but they remain tantalizing. I'll set them aside, too, and turn to a contrasting set of cases where the difficulties are more obvious. These involve situations where an individual's very existence was unambiguously contingent upon the action that allegedly "harmed" her.

If someone comes into existence under *those* circumstances, how can she complain about the harmful act, or wish that it had never happened? Surely she cannot argue that she'd have been better off without it, for if it had not happened then (unambiguously) she would not have existed at all. It seems clear that the only circumstance in which she could have a grievance against anyone would be if the circumstances of her life were so bad that her life was not worth living. To put it more bluntly, she'd have to claim that she'd have been better off never being born.[2]

If the point of criminal law is to protect citizens from certain kinds of harm (or extreme, up-close offense), it is difficult to find a legitimate role for criminal law in situations involving PGD or reproductive cloning. For exactly the reasons just outlined, it is hard to see who is harmed in such cases. Imagine that Abigail, a would-be mother, makes the decision to choose a particular embryo to implant, gestate, and bring to term, perhaps knowing that this embryo has the genetic potential for a minor disability. It appears that the resulting person has nothing to complain about, as long as she has a life worth living. *She* would not have come into existence as a fully formed human being if not for Abigail's action, even though *somebody else* might have if Abigail had chosen a different embryo. (Nor can any nonexistent person who *might* have come into existence complain about Abigail's choice, for there are no such people!) So does Abigail do nothing wrong, or at least nothing that the law should interfere with, when she chooses the "disabled embryo"—namely, the one with the genetic potential for a disability?

In a searching discussion of legal and moral issues surrounding PGD, Colin Gavaghan (2007, 74) relies heavily on such reasoning, even elevating the problem into what he calls "the Non-Identity Principle." Gavaghan (ibid., 73–76) maintains that children are not harmed when the supposedly harmful act is a necessary condition for their existence and they have at least a minimally worthwhile quality of life; further, this proposed principle holds that such acts should not be criminalized, since they do not cause harm to anybody. Thus Gavaghan combines the Millian harm principle with reasoning about nonidentity issues. Adopting his approach, we can conclude that choices relating to PGD and the selection of embryos for implantation should never be criminalized unless the resulting person's life is not worth living. Similar reasoning will apply to acts of reproductive cloning: a particular person who is created by the SCNT technique cannot have a genuine grievance about it, so long as the quality of her life is at least minimally worthwhile.

Is that an acceptable conclusion? It may seem particularly unpalatable in situations where PGD is used to select traits that would normally seem detrimental to leading a flourishing life. It is, for example, easy to imagine a case in which a deaf couple use PGD to select an embryo with the genetic potential for profound deafness. More shocking scenarios are also imaginable, though perhaps not easy to imagine *actually happening*. Here is one obvious scenario: What if a sadist deliberately chose an embryo with the potential to develop a painful disease?

Feinberg has considered similar issues in the context of wrongful life cases: civil suits in which the plaintiff seeks compensation for having been born. He defends the conclusion that somebody—let's say Barbara—is in a *harmed state* in the following class of situations: as a result of somebody else's (let's say Abigail's) wrongful act, Barbara comes into existence with a condition so bad that it would be rational to prefer nonexistence. Although there is a *wrongful act*, it is not expressed, in Feinberg's view, as a wrong to the child or a case of the child being *wronged*.

This analysis reveals many complex issues, which Feinberg attempts to tease out. Even if we accept his perspective, however, it leaves open what we should think about a case where Abigail has chosen an embryo with the genetic potential to develop a disability, such as deafness, but the resulting baby's life is not so bad that nonexistence is preferable. Consider this against Gavaghan's proposed nonidentity principle. Abigail's deliberate act brings into existence a child, Barbara, who has some kind of nontrivial disability. The facts are such that Barbara's very existence was contingent on Abigail's action. Imagine that Barbara does not regret being born and has a life that, on balance, is worth living. Once again, has Abigail done anything wrong? More important for the present purposes, has she done anything that the law ought to have deterred?

We therefore must consider two distinct classes of cases where nonidentity issues are involved:

Class 1 The birth of *this child* (Barbara, for instance) is contingent on a parent's act, and the child will not have a life worth living. According to Feinberg, the child is in a harmed state. The parent committed a wrongful act, so Feinberg suggests, but we cannot coherently claim that the child has been "wronged."

Class 2 The birth of the child is contingent on a parent's act. The child will be born with some kind of disability, but will have a life worth living.

It might be rational to confine wrongful life suits to nonidentity cases that fall into class 1, since it is difficult to see how Barbara, the child, is *worse off* in class 2 situations than she would have been if her parent had acted differently. Hence, it is hard to understand how Barbara could claim compensatory damages to restore her to a position she'd have been in but for Abigail's act. Assuming that this is correct, Barbara should not be able to sue her mother, or anyone (such as doctors) involved in her conception and birth. This seems right, but does it follow that there should be no legal limitations at all on conduct leading to these class 2 situations? More specifically, Abigail's conduct, as we've imagined it, does not fit neatly into the general doctrines of tort law, yet could such conduct nonetheless be criminalized?

In dealing with these difficult class 2 cases, Feinberg refers to the child as being in a "harmful" rather than "harmed" condition. He suggests that the parent here does something wrong, but does not harm the child. The wrong, contends Feinberg (1992, 27), relates to bringing a particular evil into the world. But isn't this drifting away from what we'd expect of a liberal theorist? Note that Feinberg's analysis assumes that bringing a disabled child into the world is indeed something to be avoided where reasonably possible. Feinberg points out that the child's harmful condition (if that is what we are going to call it) is a nongrievance evil, as the child cannot actually complain about being worse off than in some counterfactual situation where the act did not take place and they did not come into existence at all. Accordingly, the issue is raised as to whether criminalizing such an action, even where the parent's state of mind is entirely selfish or sadistic, would be too great a departure from liberal principles.

Given that this circumstance is peculiar, Feinberg responds that making an exception to the harm principle strictly applied would not be merely ad hoc. The departure would not be contrary to the humane spirit of the harm principle, and liberals can be expected to combat *harmful* conditions experienced by other human beings even when they are not what he thinks of as *harmed* conditions (i.e., even when the person's life is worth living). Though such nongrievance evils do not strictly fall under the harm principle, Feinberg suggests, their connection with human suffering makes them a special category of nongrievance evils. Where such a situation is extreme, he concludes, prohibition is legitimate (ibid., 28–31, 35–36). This does not mean that Feinberg thinks that legal prohibitions of this kind are *necessary*, or that they should be enacted *all things considered*. He points out that the most heinous cases would be unusual and that parents could be expected to avoid known risks without the need for criminal deterrence. Nevertheless, liberal principles should not stand in the way of such laws.

If this view is correct, Feinberg has specified what amounts to an extension of the harm principle. The idea is that there are situations that do not, strictly speaking, involve acts of harming but instead may legitimately be prohibited by a liberal democracy without abandoning its liberal credentials. If we accept this analysis, it is permissible for the laws of a liberal society to prohibit the use of PGD to select detrimental genetic combinations.

Is Feinberg right?

PGD and the Problem

How, then, should we regard the use of PGD to choose an embryo with the genetic potential to develop a disability, but where the child's quality of life will be worthwhile? Does the deliberate implantation of a disabled embryo harm or wrong the resulting person, and, in any event, should it be forbidden?

John Harris argues that, in fact, it *should* be forbidden even if it is technically not an act of harming. He also suggests that in any event, the embryo is harmed in terms of being caused to be born in a "disabling or hurtful condition," even if the condition turns out to be fortuitous for the child as circumstances unfold. Although Harris sees the child as having been harmed in this sense, he denies that she has been *wronged*, as long as her life is worth living, taken on balance, and she is glad to be alive (Harris 1998, 106–116).

This view has led Harris into some debate with Jonathan Glover. Glover notes that we all have some disadvantages; so have we all been harmed by our parents, who brought about our births? This conclusion is, he maintains, unacceptable (Glover 2006, 25). In response, Harris agrees that there is a sense in which our parents unavoidably harm us—but he thinks this is a trivial sense. He adds that there are powerful reasons to respect a right to found a family, and think that you benefit and do not wrong a child by bringing them into existence (so long as the child's life is worth living). On this account, you have not wronged a child if there was a low anterior probability that they would have such a grim life (Harris 2007, 93–94).

Whose side should we take in all this? Feinberg's? Gavaghan's? Glover's? That expressed by Harris? Some different view entirely? Clearly there is a difficult conceptual problem here. Our ordinary concepts of harm, harming, wronging, and so on are difficult to apply convincingly to any case where the person concerned *not only* did not exist, at least as a self-conscious or even sentient being, at the time of the conduct, *but also* would not have come to exist, at least as a fully developed human being, unless the conduct took place. On that basis, I prefer to say that the child has been neither harmed nor wronged in the nonidentity situations that I have been examining. Still, I agree with Feinberg that the parent(s) in class 1 cases, those where the child's life is so miserable as not to be worth living, have clearly committed what the law might reasonably regard as a wrongful act.

But what about cases where the child (again, let's say Barbara) is born with a disability, yet goes on to have a life worth living? Here the law would not have protected Barbara from an undesirable outcome (from her own point of view) if it had successfully forbidden the conduct that brought about her existence. At the same time, Barbara does, once born, exist in a state that Harris would describe as harmed and Feinberg as harmful. That is, she does actually suffer a disability. I prefer Feinberg's terminology on this point—the word harmful versus harmed—since it is forward looking, alluding to the disadvantages that the child will (likely) suffer in pursuing her future goals. Be that as it may, the important issue is not terminological. The issue of substance is just this: How should a liberal democracy respond to the prospect of such actions?

When defined in that way, the conceptual issue of what harming or wronging really amounts to may not be determinative. Even if the choice to implant a disabled embryo does not *harm* the embryo or the resulting child, it may still be contrary to crucial values that we cannot regard as merely optional.

A Liberal Response to the Nonidentity Problem

In chapter 1, I reported Starr's view that liberalism endorses certain values that it cannot consider neutral, such as the value of health. I concur with this, and suggest that no society—liberal or otherwise—is neutral about choices for or against the health of children. It is not surprising if liberals agree with Feinberg's attitude that it is an evil, if not exactly a harm or wrong to any affected individual, if a disabled child is brought into the world when a healthy one might have been.

Stated more broadly, a liberal society, if it is to live up to its liberal pretensions, will tolerate a wide range of moral and religious attitudes to the value of human life. It cannot be neutral, however, about a basic commitment to the lives of its new citizens going well rather than badly. Underlying this, perhaps, is the sense that any society's children are its hope for the future. Even an overpopulated society that might have good policy reasons to discourage further population growth—or seek a reduction in its overall population—has good reason to favor the flourishing and success of whatever children are actually born within its borders. In making this point, I have adopted some language from Ronald Dworkin's (1996, 141) *Freedom's Law*, even though I do not endorse Dworkin's view that such a position amounts to a way of holding human life to possess inherent importance or sanctity. Nothing so metaphysical need be involved.

Thus, it might be said that liberal societies take a perfectionist stance toward children, but if so it is a mild one that would not justify the coercive eugenic programs of the past. All the same, a liberal society's reasons for action might, after all, go beyond the protection of individual citizens from wrongs and harms.

When this consideration is added to Feinberg's observation that liberalism's humane ideals are called on when we imagine the implantation and gestation of a disabled embryo, it does appear that a liberal society has good reason to respond with hostility to some imaginable uses of PGD. If we prohibit such actions as the implantation and gestation of a disabled embryo, this is not a merely ad hoc exception to the harm principle. Indeed, it is not opposed to the considerations that led to the harm principle in the first place—for example, no otherworldly, esoteric, or inaccessible understanding of the human situation is being imposed on those who reject it. But this conclusion seems to raise a further problem. Just how far should the state insist that parents provide their children not only with love, nurturance, and some socially accepted minimum of education but also with something closer to the optimal conditions for flourishing in the society concerned?

Arguments that relate to this issue can arise in respect to many uses of enhancement technologies, but they have been most prominent in debates about the acceptability of reproductive cloning. In that context, critics often emphasize the possibility that children produced by the use of SCNT will suffer confusion about their own identity (as "twins" to people a generation older than themselves). Leon Kass (2001), in developing this thought, argues that people will always be comparing the clone's doings in life with those of the person who was cloned.

Such contentions deserve separate consideration (see chapter 4). At this stage, though, it is worth noting a problem with them insofar as they merely allege that being created through SCNT provides somebody with a less than optimal entry into society. Considerations such as these are not usually thought to be sufficient justification for the state to exercise its coercive powers. For instance, as R. Alta Charo (1999; cf. Harris 2004, 84–86) points out, the state does not forbid a woman from having a child when she is too young and impoverished to give it proper care, yet this is likely to hinder a child's life prospects significantly. Any attempt to forbid this would surely lead to a public outcry.

While modern societies place a value on their new citizens' prospects for flourishing and success, we do not typically demand that all possible efforts be made by parents to have children only under optimal circumstances or give them an optimal upbringing. Hence, any degree of perfectionism found within liberal democracies is indeed mild. Where nurture and education are concerned, the idea is not to require that children become physically and cognitively equipped to make all possible life choices, or that they be capable of making any contribution whatsoever that the society might wish or state officials might like. If Adrian and Bess encourage young Catriona to aim for a relatively undemanding and low-paid career, when she has the innate talent to be a nuclear physicist, society as a whole does not criticize this in any serious way, much less forbid it by law.

In practice, we tend to adopt a rather lenient standard: children should be nurtured, treated kindly, and brought up to be good, well-adjusted citizens who can play a productive, nonviolent role in society. Parents are expected to aim at their children's health, give them at least some generic skills, and ensure that they develop capacities that will give them a reasonable range of life choices. Furthermore, the state does not demand proof that intending parents will be able to meet even this lenient standard. In any event, public policy does not demand anything analogous to Savulescu's principle of procreative beneficence: again, this is the idea that we should, to the extent that we can, select the child with the *best* life prospects. Liberal democracies do not require a commitment to the analogous principle that parents must make whatever environmental decisions will be best to enhance the life prospects of children in their care.

None of this detracts from the following claim: the state may well take an interest in ensuring that reproductive cloning is not used for so long as it is significantly likely to lead to children with seriously debilitating or disabling congenital abnormalities. The same applies to the use of any other enhancement technology as long as it involves such risks. Thus, the state may take an interest in ensuring that PGD is not used to select embryos with the genetic potential for an identifiable and more than trivial disability. If these claims are accepted, they have a critical role to play in the formulation of public policy. Note that they do not require that PGD be used only to select embryos that meet an impossible standard of perfection; as Glover suggests, we *all* have some genetic disadvantages. Even extensive genetic testing of embryos cannot prevent this.

The state, however, may forbid the deliberate use of PGD to select an embryo with the genetic potential for, say, deafness, blindness, intellectual disability, malformed limbs, or a disease such as cystic fibrosis. These things may be forbidden even if the prospect is that the child will have a life worth living.

And yet there is something further to be said.

PGD and Detrimental Traits: Some Final Twists

Even in a liberal democracy, there may be legitimate reasons to forbid the deliberate implantation of a disabled embryo, but it does not follow that the state should always, in all the circumstances, enact legal prohibitions. Nor does it follow that the state should forbid all attempts whatsoever to bring a disabled child into the world. The possible use of PGD to create a congenitally deaf child has been the most controversial issue to date, and it is worth exploring slightly further. While liberal values do provide reasons not to tolerate such an action, it remains true that this is an instance where legal prohibition does not actually protect anybody: nobody can *complain* to the state that it allowed her to come into existence deaf (assuming that her life is one that is worth living). Moreover, the discussion up to this point shows why we might be tempted to make exceptions in some cases involving deaf parents (and similar considerations might indicate toleration in some other specific situations, such as that of dwarfism).

First, it is likely that such parents will not be acting out of sadism or malice, or even ignorance or irresponsibility, but out of a conviction that they are better placed to nurture and socialize a deaf child than one with normal hearing. If they are sufficiently immersed in (capital-D) Deaf culture, the parents might consider themselves in a position to supply a deaf child with access to a culture that they experience as rich, complex, and satisfying—and not available to those with normal hearing. They are likely to be motivated by a concern for the flourishing and success of their child, exactly like other parents.

Second, the parents' sincere beliefs cannot be dismissed as simply delusional or the product of false consciousness. On the contrary, we should show a degree of humility when we consider what they have to say to the rest of us. While they might not be in a position to assess the full richness of what they have missed out on by being cut off from the world of music, for example, the rest of us perhaps are no better placed to assess what can be substituted for it by the parents' own culture. Indeed, such considerations as these lead Gavaghan to question whether deafness is a good example to concentrate on. At the heart of the debate about deafness is a controversy about whether it really is a disability (Gavaghan 2007, 75; Wilkinson 2010, 66–68).

In a wide range of circumstances, deafness imposes significant restrictions on an individual's access to knowledge of her environment, at least compared to others with whom she may be involved in relationships of cooperation or competition. It impedes affective (i.e., emotional) communication, much of which depends not only on the ability to hear words but also to distinguish tones of voice. It cuts off the individual from much of the larger culture, including music, but also much else (Glover 2006, 23). Rather than denying that deafness is a disability at all, it would be more plausible to argue that this particular disability is one that has been addressed with great effort and creativity in modern times, to the degree that it is not always a significant barrier to a growing individual's welfare, flourishing, and success.

Where the individual's parents are deaf and immersed in Deaf culture it is even conceivable that deafness could, on balance, enhance the child's future prospects; in any event, a parent could reasonably come to that conclusion, even if other reasonable people differ. Furthermore, it is unlikely that parents would go out of their way to have a deaf child except in such circumstances. Taking all that into consideration, along with the costs of law enforcement and stigma attaching to parents if their behavior is criminalized, liberal democracies might do best as a matter of practical policy to avoid the use of criminal law in such cases.

That does not mean that the government *endorses* such conduct. It might demonstrate its lack of endorsement by placing other barriers in the way, such as refusing to fund the use of PGD for such a purpose, if it is practicable to separate this from other purposes for which it does make public funds available. Nonetheless, this example illustrates the fact that any state apparatus faces a wide range of considerations when it contemplates prohibiting any conduct, even conduct that does not fall squarely within the zone of liberal tolerance.

As an additional twist, though perhaps not the final one in this ongoing debate, there might yet be circumstances in which the state has little choice but to forbid all attempts to use PGD for the purpose of selecting and implanting disabled embryos. While it might be most prudent, all things considered, to take no action if the only likely cases involved deaf parents well positioned to socialize their children into the Deaf community, the balance of considerations would change if it turned out that some individuals sought to use PGD for sadistic or malevolent purposes, or in circumstances involving the creation of children with greatly impaired prospects of flourishing. If it became necessary to enact legislation to deal with a genuine problem of PGD being used in this way, it might not be politically realistic to make official exceptions.

But there might still be ways to deter undesirable actions without necessarily imposing criminal penalties on parents in all cases. For instance, judges might be given broad discretion to take intentions and other circumstances into account before issuing criminal convictions, or the burden of punishment and associated stigma might be designed to fall on others (such as medical practitioners) rather than parents. The main points to establish and keep in mind at this stage are, first, that liberal democracies need not be neutral when it comes to the welfare, flourishing, and prospects for success of their new citizens. In one (fairly narrow) class of cases, this can give the state an interest in enacting laws that go beyond the protection of individuals from harm. Second, though, the state should not attempt to define *optimal* circumstances for the birth and upbringing of children.

Within those relatively vague boundaries, the wisest public policy response will depend on all the circumstances facing the society concerned, and reasonable citizens will not always agree.

Conclusion

In modern liberal democracies, the least controversial reason for forbidding certain actions is that they cause or threaten direct, serious, secular, and wrongful harm to others. This narrow statement of the celebrated Millian harm principle may require some relaxation, and it needs to be supplemented by a mildly perfectionist principle relating to the potential flourishing of children. This gives us a basis for some regulation of enhancement technologies, but nothing said so far reveals a case for sweeping and ongoing prohibitions.

Nor do I believe that such a case can be established. On the contrary, as I'll argue in chapter 3, some kinds of genetic engineering may provide genuine benefits to children.

3

Genetic Engineering: What's the Harm?

While liberal democracies may have legitimate concerns about birthing new citizens with disabling phenotypic traits, they cannot be relied on to justify the prohibition of PGD for sex selection, for unimportant characteristics such as complexion or eye color, or for seemingly beneficial traits such as the potential for high intelligence. Where traits such as *these* are selected, there is no obvious cause for complaint that somebody has been born with serious obstacles to her welfare, flourishing, and success.

Similar considerations apply to reproductive cloning: we might take cognizance of safety arguments, such as those mentioned briefly in chapter 1, but liberal democracies would have little basis to forbid the use of reproductive cloning in the absence of concerns about congenital malformations arising from the process. In the absence of those concerns, and provided that the genotype of the donor of nuclear DNA does not carry the potential for seriously disabling traits, it is difficult to see why public policy should try to deter reproductive cloning. I will return (in chapter 4) to issues about the individual's autonomy, but there is, to say the least, a puzzle about the widespread political hostility to the possibility of safe reproductive cloning.

The picture changes somewhat when we consider genetic engineering. First, the technical hurdles must be acknowledged. Attempts to control the genetic potentialities of human embryos are likely to prove extraordinarily difficult in most cases. There is a risk that well-intentioned actions could lead to developmental anomalies, including some that might not be obvious until late in an individual's life or even in the lives of later individuals (Newman 2005, 205–207). Such problems may keep a technology of "safe" genetic enhancement beyond our grasp indefinitely.[1] The technical problems are truly daunting, with individual genes typically affecting multiple propensities or traits, and most traits seeming to be polygenic—depending on the interactions of many different genes (President's Council on Bioethics 2003, 39–42). Even human reproductive cloning, which is far simpler in concept, is at an impasse: there is no apparent prospect of developing a safe technology of this kind without carrying out experiments that are too risky to allow.

Safety considerations aside, our ability to engineer embryos with favored characteristics might prove to be rather limited. Such characteristics as intelligence are doubtless highly complex, and their very nature is debated (Agar 1999, 179–180)—which should be kept in mind in the practical situation that we face as we develop real-world policy. All the same, it is at least conceivable that a way will eventually be found to engineer human embryos for such things as superior cognitive and physical potentials, perhaps opening up choices that go well beyond those available with PGD, where the selection of embryos to implant is drawn from a limited number. If we ever reach this point, where is the harm in it?

One difference from the situation with PGD is that the nonidentity problem does not arise, at least not so clearly. Like Gavaghan (2007, 220), I will assume that it is coherent and reasonable for a genetically engineered person to consider herself the same individual before and after the genetic intervention took place. In that case, she may be able to complain that but for the intervention of parents and meddling medicos, she would have ended up with a different genetic potential, perhaps one more congenial to her, given her projects and values as she grows to adulthood. If all this follows, it makes sense for the state to forbid genetic modifications that harm a child's life prospects, such as by causing a physical disability or stunting her cognitive potential in some way. The most interesting cases, however, are where genetic modifications are designed to *boost* the child's prospects for flourishing, success, and welfare. Where's the harm in *that*?

Genetic Benefit

Imagine, then, that Abigail has effective control of an early embryo that she intends to have implanted in her own uterus, but only after its genetic potential has been altered in certain ways, perhaps for greater muscular strength, expected longevity, intelligence, and resistance to disease. It appears as if the child who will eventually come into existence—Belinda, this time—thereby receives a benefit. Abigail may claim with some plausibility that she is certainly not causing *harm*.

One way to dispute her claim is to challenge the whole concept of genetic benefit. This would involve arguing that what counts as a benefit is relative to an individual's physical and social environment as well as her own values and plan of life. Reliance might then be placed on Agar's (1999, 176) plausible assertion that "some sophisticated future genetics may be able to predict how a given genotype will combine with a specified environment to produce some significant capacities," although "we cannot make a similar claim in respect of life plans." That is, we are unlikely, according to Agar, to be able to engineer an embryo to become a child who wants to follow some predetermined career or way of life. This is partly because the genetic complexity involved is likely to be too great, but also because the task seems to be the nearest thing to impossible in principle. People's life plans (or simply their choices in life) are highly sensitive to environmental influences and highly specific to them.

Accordingly, so the argument might continue, the way of life that Belinda will choose in the future is unpredictable, even if we know her genetic potential. This opens up the possibility of a mismatch between her engineered genetic potential and whatever life plan she might ultimately choose, or whatever values she might develop. Bruce A. Ackerman (1980, 121–123; cf. Habermas 2003, 82–83) has contended that a mismatch between engineered abilities and inclinations could make a child miserable, if she is incapable of achieving what she wants, and would seem to give her grounds for complaint against her parents. If such a mismatch takes place, perhaps it may be said that the child was harmed.

Agar's own discussion leads him to a principle that he describes as Rawlsian in spirit, thinking of the maximin approach to distributive justice for which US political philosopher John Rawls became famous.[2] Agar's idea is that genetic engineering to modify the traits of a future child is justified only when it is done in a way that improves the child's prospects no matter what life plan she decides to pursue. In particular, he proposes, it should improve her prospects if she pursues the life plan that would be least favored by her genetic potential. The idea is to equip her for whichever life plan she might choose. The implication is that the only genetic enhancements that could ever be justified would be those capable of delivering truly general-purpose benefits.

I am sympathetic to Agar's position, but there are potential problems, and his view at least requires clarification. I will argue for an approach that is similar in spirit, and indeed may be the best interpretation of Agar's notion if it is not taken too literally. It is also similar in spirit to that urged by Fritz Allhoff (2005, 49–52), who maintains that genetic engineering is

morally acceptable in itself (leaving aside such things as utilitarian consequences or just distributions of resources) if and only if it directly or indirectly augments Rawlsian primary goods—the kinds of things we should
(rationally) all want to have, which include useful talents.

Here is the problem for proposals such as Agar's and Allhoff's if they
are applied too literally. Consider again Abigail, who is planning to have
a child, Belinda. At the moment, Belinda is just an embryo that has been
created through IVF. Soon, if everything goes well, she will become an
actual baby with a full human life ahead of her.

Abigail wishes to enhance Belinda's strength, expected longevity, intelligence, and resistance to disease. In some imaginable circumstances,
even these modifications could turn out to be disadvantageous later in
life. Consider the genetic intervention to give Belinda greater resistance to
disease and thus improved health. It just *might* turn out that Belinda will
grow up wishing that she had gained more experience of suffering disease
when she was younger. *You never know.* She might come to think that
this would have been morally improving in some way; she might think
that suffering ennobles, for example, and that therefore she is less "noble"
than she would have been if her parents had left her genome to take care
of itself. Or she might think that there is *virtue* in suffering. Alternatively,
she might come to believe that she has missed out on experiences that
would have equipped her with more understanding of her patients' needs
in the hospital where she finds work as a medical doctor.

Similar possibilities apply in terms of intelligence. Agar explores
whether intelligence is a single characteristic or whether it is a complex
of characteristics that may be mutually conflicting. According to the latter model, any attempt to boost one intelligence module, such as that for
musical ability, might reduce the individual's capacity in some other area,
such as skill in social interaction. In that event, Agar's test is clearly not
met and cannot be used to justify enhancement. Yet Agar (1999, 179–
180) suggests that it *could* be justified if intelligence turned out to be a
general characteristic explaining a wide range of performances.

For the sake of argument, let's conceive of intelligence as a cognitive
capacity—or a complex of related, nonconflicting capacities—to reason
quickly and accurately, understand difficult concepts, and solve problems.
We may stipulate, further, that improvements in the genetic potential to
develop this capacity do *not* cause any loss of genetic potential in other
cognitive capacities (or physical ones, if it comes to that). Even on these
assumptions, which seem favorable to the proenhancement case, no *guarantee* can be given that high intelligence will improve the prospects of all

children, irrespective of the life plans they ultimately adopt, values they develop, or circumstances in which they find themselves.

It is quite conceivable that Belinda will grow up wishing she were less intelligent. For example, her high intelligence might lead her to skeptical thoughts about her society's traditional religious and moral beliefs. Some people might take pride in such skepticism, but Belinda might be attracted to a life of religious devotion and conventional morality, and her skeptical thoughts might be unwelcome to her, perhaps even a source of anguish. Or her intelligence might attract social pressure for her to enter a learned profession that she finds unattractive, or that she tries and dislikes. Habermas takes this sort of line, observing that the effects of high intelligence can play out in different ways. Not being able to forget can sometimes be a curse, he says, and the lack of selectivity of memory can be counterproductive if we are overloaded with too much data (Habermas 2003, 85–86).

As for Belinda's superior strength, this will make it easier for her to excel in many physical activities, but it may cause her regrets if she comes to value the kind of psychological suffering that frequently accompanies the experiences of failure. Finally, even a long life can be a curse in some circumstances. Belinda might ultimately outlive her loved ones by many years. As a robust centenarian, she might find herself wishing she had died twenty years before—when she was still in her eighties.

These objections, which we could label Habermasian, seem to show that Agar's Rawlsian principle cannot justify *any* form of genetic modification of our children. Habermasian objections, though, surely prove too much—for instance, they could militate against such environmental interventions as vaccination (as, you *never know*). But that approach to the world suggests that we can never act benevolently in any way whatsoever to anyone, since there is always some risk that our action will end up harming someone. Perhaps I am absolved of the risk if the person concerned expressly accepts a gift, but in many cases we must act for people's benefit without obtaining permission or acceptance. When dealing with young children, we must certainly act as we see best to benefit them, at least in matters that are beyond their understanding. And yet nothing we do is ever guaranteed to assist another person's life prospects. In imaginable situations, even the provision of Rawlsian primary goods might be detrimental to them.

I suggest that we clarify the test so that it does not require parents to be perfect predictors of the future. With that clarification, the Habermasian type of objection evidently fails. Almost anything we do might have terrible though unpredictable consequences, but that does not prevent us from forming loving relationships, treating others with kindness, or giving donations to Oxfam. In general, these will have beneficial results and no particular bad results can be foreseen. You never know, of course, but that is not the point. Thus, we should not have to argue that a general-purpose benefit such as intelligence should benefit any possible life plan. It should be enough that it offers a wide range of improved life prospects, with no predictable downside. By contrast, we might hesitate if intelligence is more modular, and if enhancing one "module," such as mathematical ability, would ipso facto require reduction in another, such as skill in understanding verbal nuances. We would not go ahead with enhancement in such a case.

That, I think, is the spirit of Agar's proposal. I would go further, however. While we might worry if an enhancement in one area produced a diminution in another, we are surely entitled to take account of the relative size of these effects. What if we could produce a substantial enhancement of mathematical ability for only a slight reduction of skill in understanding verbal nuance? It's not obvious that we should refrain, as long as we can be confident overall that we are improving the child's life prospects. Care needs to be taken, because there could be many scenarios, perhaps some with collective outcomes that we want to avoid (would a sharp rise in average mathematical ability across an entire society really be wanted if the effect was to reduce the number of people with genius-level verbal skills?). Nonetheless, there are clearly some cases where even a clarified version of Agar's approach is too demanding, and we can say with confidence that a child's overall prospects have been improved even if they have suffered *some* diminution in one or two respects. That, in my view, is closer to our thinking with environmental interventions, where trade-offs always have to be made.

Several more points can be made here to put Habermasian objections in their proper perspective. First, the argument that any characteristic at all can be disadvantageous in *some* imaginable circumstances has a bizarre corollary: any characteristic can be *advantageous* in some imaginable circumstances. The characteristic might be physical weakness, premature aging, low intelligence, or susceptibility to disease—these could all have advantages in some situation or other. Habermas (2003, 86) states, correctly, that even "a mild physical handicap" might end up being advantageous. To this, Harris (2007, 142) responds that parents cannot even know whether a *severe* handicap might end up being advantageous. Does this demonstrate that the playing field is level, that someone who engineered her child for such traits as physical weakness, premature aging, low intelligence, and susceptibility to disease would be morally equivalent to someone who engineered her child for strength, intelligence, and so on?

An affirmative answer to this question seems unlikely when we consider what most of us would think of a parent who used *environmental* means (such as poor nutrition, a dull range of early experiences, etc.) in an effort to ensure that her child grew up to be frail, stupid, sickly, and short lived—or with the mild handicap referred to by Habermas. To push this further, I expect that most people would take a dim view of parents who made no efforts, one way or the other, in respect of, say, their child's resistance or susceptibility to disease.

Note that some ingenuity was required to produce counterexamples to the idea that strength, resistance to disease, intelligence (as conceived here), and a natural potential for longevity are good characteristics to have. That should be a warning: the very ingenuity involved in developing counterexamples indicates that even in a situation where someone, on balance, wishes they were not so healthy or not so intelligent (for instance), there is something about health and intelligence, considered in themselves, that seems to resist being disparaged. What's more, this should not be surprising.

Notwithstanding seeming counterexamples, there are many real-life situations in which we might reasonably desire to retain, or boost, such characteristics as our health and intelligence. For a start, there is a certain joy in functioning healthily, solving the problems that we encounter in life, and obtaining a more accurate apprehension of our situations and the larger reality in which we find ourselves. Moreover, health is instrumentally valuable in a wide variety of situations that human beings encounter in many kinds of environments. Health is a form of power: possession of good health can assist us to function and act more effectively than if we were unhealthy. It can help us to achieve a wide range of personal goals as well as to act in ways that are beneficial to those around us, our society as a whole, and other beings whose interests we wish to advance.

Much the same can be said of intelligence. Possession of intelligence tends to lead to greater knowledge and a better understanding of our particular situations, along with the ability to master techniques of inquiry that will enable us to increase our knowledge and understanding still further. If knowledge and understanding are valued for their own sake, intelligence gives us a more realistic prospect of obtaining them. In any event, the possession of intelligence can enlarge our life possibilities, enabling us to carry out tasks that require understanding and problem-solving abilities. If we are intelligent, we will be better equipped to achieve a wide range of possible goals, since we are more likely to act from a genuine knowledge of what means will be effective. Similarly, we will be better equipped in a wide range of cases to take actions that contribute to the well-being of others.

Intelligence may not be a *moral* virtue—a desirable disposition of character—but it is a human excellence of value to almost anyone who possesses it. In *What Sort of People Should There Be?* Glover goes a step further and writes of the long-term possibility that we may eventually come up against a limit to our intellectual capacity for understanding of the universe we live in. We might then want to enhance ourselves to transcend this limitation, in which case our descendants might think we made a decision that was an "escape from a kind of claustrophobia" (Glover 1984, 179–181).

This analysis suggests that Abigail has a strong rationale to support her claim that she is actually benefiting Belinda rather than harming her. Genetic potentials for strength, longevity, intelligence, and disease resistance are good things to be born with, even if they can lead to some bad consequences for some people in some imaginable circumstances. Genetically engineering a child to give her such characteristics should be thought of as bestowing a good upon her, in much the same way as one bestows a good on a child by bringing her up in an intellectually stimulating environment, aiding her physical growth with exercise and good nutrition, and ensuring that she receives vaccinations against dangerous diseases. We can concede that no benefits will turn out to be instrumentally good for every individual in every conceivable situation, without denying that they are, by all reasonable human standards, benefits.

This line of reasoning appears to apply as well to the enhancement of genetic potentials for affective understanding (or "emotional intelligence") and whichever moral virtues are genuinely beneficial to those who have them—courage, perhaps, but perhaps also trustworthiness, kindness, and generosity. For the purposes of this chapter, I set aside the question of whether attempts to use genetic interventions to mold a child's personality are threats to her autonomy. I'll take up that issue in chapter 4, but for now such enhancement of virtuous dispositions can be classified as a benefit rather than a harm. Similar logic applies, perhaps to a lesser extent, to a wide range of more specific characteristics, such as musical ability and sharp eyesight. For most people, in most realistically foreseeable circumstances, these characteristics are good things to have and bad things to lack.

To say that health and intelligence, for example, are general-purpose benefits is not to make the strong claim that they are beneficial for every possible kind of life. Nothing could meet such a test. Nor is it to say that they supply a guarantee of happiness, in any sense of the word happy. Nothing guarantees *that*, either—not Rawlsian primary goods, and not even a full share of the most uncontroversial moral virtues. But there can be characteristics that tend to improve rather than harm individuals' prospects, while also equipping them to make contributions to social productivity and the interests of those around them.

Reasoning such as this explains why we do not normally accuse parents of harming their children, or putting their happiness at risk, when they take *other* actions that are likely to be effective in protecting their children's health or improving their prospects of growing up to be intelligent adults. For example, we do not criticize parents who arrange for their children to be vaccinated against infectious diseases. Nor do we normally criticize parents who expose their children to experiences that are thought likely to stimulate the development of their intelligence. It would be a perverse parent who declined to give her child a vaccination against polio or read to her in bed on the ground that the child just *might* grow up to regret the parent's well-meaning action.

Is It a Red Queen Race?

But perhaps it could be argued that the advantages to individual children of general-purpose enhancements are merely positional goods that would cancel out if widely distributed. Those seeking an advantage for their children might then discover that all their efforts merely made them stand still, compared to others. If this is the correct picture, attempts at enhancing genetic potential would be collectively self-defeating.

Indeed, if everyone received the "advantages," there might be a downside. After all, every action has risks. Some genetic interventions might go wrong, producing unwanted and unforeseen outcomes, such as if some individuals suffered brain damage as a side effect of genetic interventions. If this happened, disadvantaging the individuals concerned, while no one ended up obtaining any real benefit, the result, assessed at the collective level, would be an overall loss. Even if nothing like this went wrong, the effort would produce a massive waste of resources. Such a pessimistic analysis might provide a reason for prohibiting attempts at genetic engineering to enhance human potential (Glannon 2001, 99).

Once we regard even general-purpose enhancements as merely positional goods, the provision of these enhancements to our children becomes a race with no winner. If I utilize genetic engineering to ensure that my child has an unusually high potential for intelligence, say, this will be futile if other parents are doing the same, and if intelligence is merely a positional good. In chapter 2 of Lewis Carroll's (2009) *Through the Looking-Glass and What Alice Found There*, the Red Queen explains to Alice that "*here*, you see, it takes all the running *you* can do, to keep in the same place." When it comes to enhanced abilities, all the efforts of parents and others will be equally futile, bringing no good to children, if we assume that the enhancements are merely positional goods. Successive generations of children might be more and more intelligent as well as healthy, for example, and yet no one will actually be better off.

Many commentators have picked up on this theme. Buchanan and his coauthors (2000, 185) use height as an example of a positionally advantageous yet collectively self-defeating intervention. Peter Singer (2009, 282–283) uses the same example, as does Francis Fukuyama (2002, 95)—though the latter also tries to make the point with the case of intelligence: if everyone has more intelligent children, nothing is gained, except perhaps economic productivity; by analogy, an intellectual arms race to get into Harvard University does not create more spaces at Harvard. Similarly, Mehlman (2003, 89) asks what good it is if everyone has twenty points added to their IQ or if everyone can run five miles per hour faster—except for the gain to the providers of enhancements and perhaps the availability of more exciting sports events.

But the argument is flawed. It relies on a highly controversial picture of solely positional advantages that cancel out if they become widely available. The first problem here is that much more needs to be said before we set aside the possibility that some goods are objective, and that these might include such things as health, intelligence, and beauty. It might not be intellectually fashionable to imagine that the world would be better if it contained a much higher proportion of, say, physically beautiful people, and it certainly might not be fashionable to think of human beauty as an objective good. I find neither idea intuitively likely. Still, it is not obvious why people should be prevented from acting on the basis of such ideas, even if they are false—in the absence of more obvious harms than we've identified so far, what business do we have in trying to stop them?

Even if we set all this aside, the idea that phenotypic characteristics provide merely positional advantages is more plausible with some characteristics than others.[3] Perhaps it is reasonably plausible with such characteristics as height and beauty, but even beauty may not be a clear-cut illustration. At least *some* standards of beauty may reflect underlying health and physical robustness, and enhancing beauty might actually require an underlying enhancement of the genetic potentials for these things.

In any event, we can easily imagine genetically engineered individuals using their talents to benefit everybody through their achievements in art, music, science, technology, athletics, and so on, with an overall increase in social wealth and productivity (Resnik 1993, 35; Bailey 2005, 230). In a similar vein, Ronald M. Green (2007, 31) mentions the social benefits if an airline pilot with improved vision and quick reactions avoids an accident, an "Einstein-like" thinker gains important cosmological insights, or a neurosurgeon with exceptionally steady hands saves a patient's life.

The authors of *From Chance to Choice* hold that many things could be instrumentally advantageous or intrinsically beneficial, even if everyone had them, such as better immunity or resistance to memory loss. They emphasize the extensive range of ways in which parents currently try to develop their children's abilities as well as their prudence and moral virtue—all of which might be pursued through genetic and environmental interventions. Memory would be a general-purpose benefit, improving the capacity to pursue almost any plan of life, just as the loss of sight is damaging to almost any life plan (Buchanan et al. 2000, 156–159, 167–168, 186).

As for the analogy with limited positions at Harvard, surely we can turn this around. Glover notes that going to a university may produce many benefits beyond mere positional advantage: you may meet wonderful people, become absorbed in ideas and great literature, and so on. Likewise, if your parents give you intelligence, energy, and creativity for the bad reason of positional advantage, you may still not regret their choice. More speculatively, Glover suggests, we may one day need to enhance human intelligence or it will be the end of science; it may turn out that some concepts needed to understand the universe in all its complexity are beyond our current intellectual limitations (Glover 2006, 80–81, 99–101; cf. Glover 1984, 179–181). Furthermore, we often find intrinsic value in such things as studying philosophy, composing music, and doing charitable work. If so, Ronald Lindsay (2005, 21–22) observes dryly, it is difficult to find a basis to prohibit or restrict enhancements that would enable individuals to do these things more effectively.

Among the critics of enhancement technologies, Mehlman and Botkin concede that some genetic interventions, such as for greater strength or stamina, would "arguably" improve the quality of life of those possessing them. They also acknowledge that there might be substantial social benefits if dramatic enhancements could be developed. For example, humans with enhanced physical capabilities might be able to explore regions that are off-limits to most of us, such as outer space and the deep sea, while enhanced cognitive abilities "could stimulate new arts and industries"

(Mehlman and Botkin 1998, 53, 121). But these concessions are made in an unnecessarily grudging tone. Many benefits would be widely valued for their own sake. They would bring not only a positional advantage in competing for scarce resources but also the creation of nonphysical products that do not diminish as they are shared, such as knowledge and understanding.

Though little may depend on this, even height may not be a perfect instance of merely positional advantage. Agar reminds us that there are circumstances in which height offers an advantage that is independent of whether anyone else possesses it, such as reaching for apples on trees. While this example may seem trivial, Agar adds (quite plausibly) that intelligence is of independent value to those who possess it. Even if it is sometimes sought for positional advantage, it brings independent benefit, such as greater understanding of the world. Agar also introduces a narrower concept, however: a characteristic that is of advantage only in a winner-take-all competition. He points to a contrast between extreme athletic ability—possibly including such things as the extraordinary height of a professional basketball player—and extremely high intelligence. The difference is that the former can produce rich rewards only for a few people. Using this approach, pursuit of winner-take-all advantages for children could create a futile race between rival parents. This would waste resources, produce ruinous disappointments, and encourage parents to channel their children's pursuits into a narrow range of options (Agar 2004, 126–131).

Unfortunately, it will not be easy for lawmakers or regulatory agencies to identify winner-take-all advantages. For instance, how should the law distinguish between extreme athletic ability that is useful only for a narrow range of life plans and the "ordinary" sort of athletic ability? Unusual strength, speed, balance, hand-eye coordination, and similar physical capacities can be useful for many pursuits outside the ruthlessly competitive ethos of elite professional sports, although perhaps less so now than in earlier times; the nonpositional value of these characteristics might be greater for warriors or Stone Age hunters than for twenty-first-century office workers. Even now, however, it is easy to think of a large range of occupations or recreational pursuits where these characteristics might provide either a relatively benign competitive advantage or an advantage that is independent of competition with others.

Moreover, there is another problem: What do we say if a parent succeeds in manipulating her child's genome to produce the potential for outstanding intelligence *and* outstanding athletic ability? In such a case, it does not appear that the parent's intention is to force the child into narrow pursuits but rather to give her the best possible base of natural talent as the child shapes her life plan. If we apply a dominant purpose test, as Agar proposes, we may be confronted with the need to make difficult judgments that include not only what modifications were made to a child's DNA but also what other modifications could have been made, yet were not.

On the other hand, Agar's concern about winner-take-all advantages provokes an unsettling image. It is conceivable that some parents in fact could enter into a destructive Red Queen race in which they commission children with more and more extreme body sizes and shapes, specialized metabolisms, and hypertrophied organs such as lungs and heart—all in an obsessive effort to fit their offspring for elite competition in specific sports. This kind of race might have adverse effects on the children's health, shorten their life expectancy, perhaps hinder them in some activities outside the area targeted by the parents, and raise strong suspicions about the willingness of parents to provide their children with balanced life experiences.

If taken to an extreme, such parental actions could be reasonably seen as going against their children's interests and those of other children with similarly minded parents, and so be discouraged by the state. In some circumstances, it might be relatively easy to draw a line; for instance, extreme genetic redesign might threaten to reduce a child's life expectancy, or at least create the threat of health problems and off-field social inconveniences. Think of possible attempts to engineer eight- or nine-foot-tall superathletes. Perhaps legal regulation could draw a blurred line, using standard legal terminology that allows for judgments of reasonableness and dominant purpose. Wherever a case is not dramatic and clear-cut, though, I suggest that liberal tolerance be exercised in our liberal democracies.

Finally, we should question the idea of a society or social milieu as a zero-sum game. Though all societies contain an element of competition for resources, sexual success, and status, that is not the whole story. Allen Buchanan has gone the furthest in emphasizing the mutual advantages that could arise from enhanced human capacities. He discusses a powerful consideration that has generally been overlooked in the human enhancement debate: many capacities become more beneficial to each individual as more individuals come to enjoy them. This is obviously correct. Literacy, for one, is more valuable to an individual in a setting where many or most other people are literate. A highly musical person will thrive in a milieu of other highly musical people, and so on.

Buchanan argues that enhancements of cognitive capacities, duration of life, health near the end of life (with compression of the period of morbidity and disability near life's end), and disease resistance would lift human productivity as well as enlarge the possibilities of human well-being. In that respect, they would function much like what he calls the "great historical enhancements" of the past, such as literacy, numeracy, agriculture, science, legal and political institutions, and computerization. Such advances create synergies or network effects, with greater advantages to individuals the more individuals are involved (Buchanan 2011a, 116–120; Buchanan 2011b, 36–50). This claim requires some caution— the immediate effect of agriculture, say, may well have been a lower standard of living and lessened life expectancy for most people. Nonetheless, Buchanan is surely correct that changes affecting many people need not simply leave everyone's welfare as it was (or reduced overall). Productive synergies can emerge as more individuals in a society find their individual capacities enhanced, whether by new technologies or new institutional arrangements.

An Obligation to Enhance Our Children?

A related set of questions is whether, under certain conditions, it might make sense to condemn parents who do *not* take steps to enhance their children's genetic potentials, and whether it might, in some circumstances, be appropriate for parents to claim a positive right to the resources required. The plausibility of such claims should be reduced by an approach of avoiding superlative moral ambitions. While we expect parents to nurture children, develop their talents, provide them with knowledge of the world, and so forth, we do not usually require that they ensure that their children have the widest-possible range of options in life.

Thus, the test is not an extreme kind of procreative and parental perfectionism but instead a more lenient concern with the ability of children to lead lives that will go well. Glover sums up the idea by saying that we owe children a decent chance of a happy life. He interprets this in a lenient way—the child must not be placed at risk of having a life not worth living—though he seems to believe that we should remove obstacles to flourishing where this is not too burdensome (Glover 2006, 54–63). Given the wide range of parental choices in socializing and educating children within a liberal society, no narrow perfectionist standard should be required when it comes to genetic interventions. As Gregory E. Pence (2000, 98) observes, the deplorable experience of the US eugenics movement in the early decades of the twentieth century might suggest that we should never make parents feel that they are morally obliged to act in certain ways with respect to the genetic potentials of their children.

In any situations that are actually foreseeable, Pence might be offering good advice. Yet *could* there be circumstances—perhaps far-fetched or unusual ones—that would justify a different response? In some conceivable instances, might a minimal level of intervention be necessary in order for children to have acceptable life prospects? In principle, an affirmative answer appears to be required. And might this "minimal" level involve more than genetic therapy to prevent disease?

Buchanan and his collaborators give the example of a future society in which a certain level of mathematical ability is needed for all but the worst jobs. In those circumstances, we might feel that a genetic potential for only a low level of mathematical ability is something that ought to be rectified. Intervention here might be viewed as morally required not only for the purpose of treating disease but also for the alteration of an important nondisease trait (Buchanan et al. 2000, 71–74). The case would be stronger if the technology of a future society could achieve this without great personal difficulty for the parents, and if it did not burden them with an open-ended obligation that could destroy all their discretion. In sufficiently compelling situations, then, even liberal-minded people could see the refusal to make a nontherapeutic genetic intervention as harmful to the prospects of the child concerned.

I have written to this point of what the parents ought to do *morally*, but in principle it also seems that the state could impose a mandatory requirement—a *legal* obligation. Whether this would be the best policy in all circumstances is, of course, an additional issue. There would be obvious problems if parents were punished and stigmatized for well-intentioned decisions (perhaps based on religious faith or sincere moral views). Moreover, it would be unfair for the state to compel certain behavior unless resources were available for everybody to comply. This, in turn, would require public funding for those who could not otherwise afford access to enhancement technologies.

The conclusion that there could be moral and even legal obligations on parents to use enhancement technologies might worry some liberal philosophers who are not otherwise antagonistic to the idea of human enhancement. Their scruples would be mistaken, though. Under *current* or *foreseeable* conditions, the idea of any such obligations seems implausible. It might be reasonable within day-to-day contexts to dismiss any idea of moral and legal obligations to use enhancement technologies. But is the notion so implausible in contexts of critical reflection, where we ask about a much wider range of possible societies, including those that are far more technologically advanced than ours? If, in current or foreseeable circumstances, a moral duty to enhance were identified and then backed by legal sanctions, this would transgress our well-justified notions of reproductive freedom. There is no prospect of this changing any time soon, if ever, yet it does not follow that we would have the same reasons in any scenarios whatsoever that might conceivably arise in the future.

Accordingly, I part company with Agar on this point. Agar (2004, 84–87) searches for a principle that will ensure parents can never be placed under a legal obligation to intervene in their children's genes in a way that goes beyond the treatment or prevention of disease. His example is a tweak to the 5-HTTLPR region of the serotonin transporter gene. The modification would provide the "long version" of the relevant DNA sequence, which is associated with a more upbeat personality.

Agar does not argue on the basis of indirect or intangible harms to society (my topic for chapter 6). Rather, he claims with no real explanation that we should resist the idea of an obligation to provide 5-HTTLPR therapy, even if it is free, painless, instantaneous, and generally imposes no burdens on the parents or anyone else. We therefore are imagining circumstances where, among other things, none of the current pain, inconvenience, and hardship of using IVF has any relevance. In the absence of some deeper justification for this position, I see no reason in principle why a future society sufficiently different from our own could *not* justifiably impose some minimum obligations on parents to employ genetic manipulation to enhance their children's functioning. Even if the design of an upbeat personality does not provide a good example, perhaps because it has an individual or collective downside, there are others that might be more plausible, such as a threshold of mathematical ability. Would it *never* be legitimate to enact laws requiring that this be genetically tweaked? I submit that in some imaginable circumstances it could legitimately be made compulsory.

Is this an illiberal viewpoint? Not at all; let me explain.

An End to Liberalism?

Liberal democracies should not impose controversial religious or esoteric moralities for their own sake. Hence, the legislature should not require that I attend certain church services or avoid others, or that I must publicly profess certain spiritual doctrines. It should not impose a teaching that homosexual conduct is a sin or that we ought to fast at certain times of the year. Many people hold to these beliefs for religious reasons, but in some cases they simply believe that certain actions—again, homosexual conduct supplies a good example—are "wrong." The citizens of liberal democracies are entitled to such beliefs, even though they are not entitled to have the state take their side and impose the beliefs by coercive power.

But the state can take action to protect the worldly interests of its citizens, including those of children. While it should not generally take a stance on the truth or falsity of a religion, or any esoteric moral system associated with a religion, it need not tolerate, say, human sacrifice. In such a case, however, we should observe that the state does *not* reason as follows: "The doctrines of this neo-Aztec cult are false; therefore, we can ban its associated practices." The thinking is much more direct: the state acts to protect the ordinary interests of citizens in preserving their own lives, and so bans all murders, including those that take the form of human sacrifices. While people may be entitled even to extreme religious and moral beliefs, such as the belief that human sacrifice is commanded by the gods or required to preserve the cosmic order, they are not entitled to act on them when it leads them to harm others.

In a liberal democracy, we should expect the state to exercise reluctance when it comes to solving controversial moral issues. It should lean toward a toleration of diversity, and in any event it can never guarantee that its actions are morally correct by some transcendent standard grounded in a true comprehensive worldview, whatever it might turn out to be. Instead, the state confines itself to the secular welfare of its citizens and any others who the citizens choose to assist (the poor of other countries, for instance). If its decisions turn out to conflict with the will of a real deity or some metaphysical principle, that is a risk that it has to take. Furthermore, the state and its officials should not usually claim to know better than autonomous adults or mature minors what is in their own interests. To a large extent, they should avoid paternalism, leaving sufficiently mature citizens a wide discretion to pursue their own happiness in their own ways.

This is all familiar enough, but notice how the situation is different when it comes to children. Of course, liberal democracies, like other societies, grant parents extensive legal rights over their children, permitting them to make judgments about the children's spiritual as well as secular interests. Although that arrangement is fairly uncontroversial, we should note that parental rights over children ultimately exist for the sake of the children themselves; they are not a benefit for parents (Blackford 2012, 142–144).

In many cases, or perhaps most, this causes no great problems; generally speaking, parents have the best interests of their children at heart, and act accordingly. At the same time, we don't require that children receive an optimal upbringing—if that could ever be defined with suitable objectivity. The state does, though, impose a requirement that parents provide their children with certain benefits that are considered essential to their welfare, such as health care and basic schooling. Parents are left free to act as they wish *above* these minimal standards, yet they are required by law not to fall *below* them. In all, the system works with a fair bit of give and take. The law nonetheless may sometimes collide with the religious and moral views of certain parents. The latter might think, for example, that it is morally incumbent on them to keep their children ignorant of certain matters that are prescribed in the state's secular educational standards.

It should not be surprising at this point that the interests of children could require the use of enhancement technologies in at least some circumstances, even if these are rather far-fetched, science-fictional ones. Dov Fox (2007, 22–23) is correct to emphasize the possible analogy between environmental interventions such as in the form of education and the provision of some genetic enhancements. If a future society has changed so much that some minimum level of enhancement is, in practice, now needed for children to have a reasonable range of life choices within the society, then there is nothing illiberal about requiring parents to provide that minimum level. If resources are not available to all parents for them to do so, they should be provided through the tax-transfer system. Above the minimum, parents should still have extensive freedom to do as they think best.

Whether any specific obligations to use enhancement technologies could be defended would depend on their consequences for all concerned. In the situations imagined by Agar, where the technology operates seamlessly, is used by almost everyone, and imposes no burdens on its users, there might be significant hurdles in life for the rare person whose genetic makeup has *not* been modified in certain specific ways. Given those circumstances, it actually might be justifiable to impose a legal obligation. That might not be the best policy when all things are considered—perhaps there might be reasons to accommodate the passionately held beliefs of conscientious objectors—but liberal principles do not rule it out.

The important conclusion, then, is that it is *not* the case, in all imaginable scenarios, that people have, or ought to have, a right not to use new reproductive technologies should "they find the enterprise offensive or inimical to their own reproductive project" (Robertson 1994, 172). It is valuable, no doubt, to ascribe such a "right" in current circumstances, and (I suggest) in the circumstances of any realistic scenario for the medium-term future. But the presence of such a right is not a timeless moral truth, applicable even to societies of the future whose conditions may, for all we know, depart radically from our own. Again, we are talking about a society so technologically advanced that using genetic enhancement creates no physical, social, or economic hardship, though it may be against some people's religious or moral beliefs.

Once again, there is nothing illiberal about this. The state is acting to provide a minimal level of protection for the interests of children. It is not imposing any religious or esoteric moral views for their own sake. There may be an indirect effect on parents whose religious or moral views are in opposition to the state's standards, and that is doubtless unfortunate. But the state can never guarantee that there will be no collisions between the conscientious scruples of some parents and its own secular standards for protecting children's welfare.

Conclusion

Any action at all can produce unfortunate outcomes that could not have been specifically predicted, but that does not prevent us from acting in the ways that are most likely to benefit ourselves or others, given the information available to us at the time. Habermasian arguments can be developed to the effect that any efforts at genetic engineering could be disastrous, since there is always *some* possible scenario in which they could prove detrimental to the life plans of the person affected. Yet that should not prevent us going ahead with interventions that are most likely, when viewed in prospect, to be useful. Thus, we can identify some kinds of genetic engineering as providing genuine benefits to children.

It is possible to defend an even stronger claim: that there could be future situations in which the state has legitimate reasons to make certain genetic modifications compulsory. Whether this would be the best policy, all things considered, is another matter. Depending on the precise circumstances, it might be advisable to accommodate the sincere religious or moral commitments of parents who object to using enhancement technologies. Still, there is nothing fundamentally illiberal about imposing minimum standards that protect children's secular welfare.

If current societies change sufficiently over future generations, the deeper values that underlie liberalism may indeed support some positive pressure to employ enhancement technologies. If a certain level of enhancement is necessary for children to have a reasonable range of life choices within the productive networks of the future, then future liberal democracies will and should set standards. Intuitions to the contrary seem to be based on the mistake of assuming that specific areas of parental discretion that are justified under current conditions are justified in some *absolute* sense—and must be continued even in circumstances that are radically different from those that give them their justification in the first place.

Up to this point, I have alluded only in passing to one important group of arguments—those related to alleged threats to autonomy from human enhancement. These arguments have attracted sufficient interest and support to merit consideration in a separate chapter. This, accordingly, will be the subject of chapter 4.

4

A Threat to Autonomy?

Some critics of enhancement technologies, particularly of human reproductive cloning and genetic engineering, point to what they characterize as threats to autonomy. Their claims take a variety of forms, but the arguments are usually developed in such a way as to suggest one or both of two things: there will be a threat to the autonomy of the resulting child and/or a threat to their society's liberal political values.

The most common sort of argument evidently relies on an intuition that if I knew my personality had been shaped by conscious decisions about my genetics as well as my early education, I would see this as undermining my autonomy (in some sense). George J. Annas, however, considers a wider range of possibilities: he opposes parental interventions to give their children higher levels of "memory, immunity, strength, and other characteristics that some existing humans have." These interventions would, so he thinks, put too much control in the hands of parents and violate children's autonomy (Annas 2005, 41).

Various related contentions appear in the philosophical literature, but the best known and most elaborate is that of Habermas. Habermas agrees with other critics of genetic modification that it poses a threat to the autonomy of children, but adds the further twist that this, in turn, is a threat to the values of modern liberal societies and apparently liberalism's ongoing viability. Somerville, who cites and relies on Habermas's views, sums up much of what is at stake here by claiming that designed children will not be equal to the designer. The concern is greater, she thinks, if the intervention is meant to design somebody's psychological characteristics. For Somerville (2007, 214–215), genetic interventions in human potentiality are a form of tyranny over the future, exercised by those in the present.

How should we respond to all this? I will maintain that despite the psychological attraction, at least for some, of intuitions about violated or damaged autonomy, the arguments have little merit. Some residue of concern does remain, even after the claims are examined, and this may justify an element of caution, especially if direct attempts are made to alter psychological characteristics. Nonetheless, the critics of genetic intervention whom I discuss in this chapter face serious problems. Not the least of these is the need to identify a type of autonomy that is genuinely applicable to "ordinary" people, while also being genuinely threatened in the case of genetically engineered people. As I'll explain, that task is far from straightforward.

Concepts of Autonomy

Making Our Own Decisions

The word autonomy can be used in various senses, some of which relate to political ideas of freedom or liberty, others to metaphysical ideas of free will, and still others to psychological concepts of self-reflection. Before we go any further, it is useful to identify the word's most important senses, and understand why no argument against genetic interventions follows directly from any of them.

In one important sense of autonomy, it refers to the ability to make our own decisions, particularly about significant matters to do with how we live our lives. A moral or political principle of respect for others' autonomy, using the word in this sense, would require a degree of deference to the self-regarding choices of other competent people, even if we believe their choices are imprudent. This is the kind of autonomy that Glover discusses extensively in *Causing Death and Saving Lives*, where he suggests that respect for autonomy can ground an objection to killing other persons, over and above any utilitarian, or similar, consequences. On an account such as this, when we respect another individual's autonomy we give priority to her decisions about her own future, accepting her present outlook on such matters, even if we can reliably predict it will change. Glover (1977, 78–80) believes that many of us would not be prepared to surrender making the major decisions in our lives even if we gained a great increase in other satisfactions. Though there is always some risk in predicting others' responses to such thought experiments, this strikes me as persuasive.

Importantly, though, Glover identifies three necessary conditions before we can meaningfully respect an individual's autonomy in this sense. First, the individual concerned must actually exist. Second, she must have developed to a point of being able to have relevant desires about her own future life. Third, she must *actually* have such desires—that is, we can violate a person's autonomy only if our action conflicts with desires that she really has (ibid., 77).

There might be some exceptions to these requirements. Consider an individual who no longer possesses the faculty of autonomy, but who previously provided an advance directive as to how she should be treated in her current condition. It might be said that her autonomy is being violated if the advance directive is not followed. Yet that situation is remote from the situation of an embryo, which has never had the faculty to begin with.

What should we say about the genetic modification of embryos, which have not reached anything like the point of development where they can have desires? Embryos are simply not the kind of thing that can exercise autonomy. At first glance, that looks like the end of the story: an embryo's autonomy cannot be violated because it has none to violate! This reasoning applies to health-related genetic modifications, aimed at eliminating the potential for disease, and any other attempts at genetic engineering, irrespective of what phenotypic outcome is aimed at. Moreover, it applies to decisions about implanting particular embryos—and to reproductive cloning. In the latter case, at the time the SCNT technique is used, there is no being in existence whose autonomy could be violated.

This is not necessarily fatal to the assertions made by Annas, Somerville, Habermas, and other commentators, none of whom relies explicitly on a claim that human embryos, gametes, or cell nuclei possess anything similar to desires or life plans. At this stage, I wish to do no more than clarify that the argument is *not*—or at least should not be—about whether genetic engineering (or any other use of enhancement technologies) would violate or infringe on the autonomy of, say, embryos. It is clear that any plausible claim based on concepts of autonomy must work in some other way.

Wilkinson (2010, 49) has such difficulties in mind when he distinguishes between a *failure to respect autonomy* (where an autonomous person's entitlement to make her own decisions is overridden or violated) and a *failure to promote autonomy*. In the latter case, we might be concerned with how the genetic engineering of an embryo could affect the person who the embryo will become, some years in the future. Will that person end up being less autonomous?

Surely that is highly problematic. The person who exists in the future will doubtless have various desires, including desires that relate to important matters in her life. There is no reason to doubt that genetically engineered people will have such desires, like everybody else, and nor is there any entailment between being genetically engineered *now* and having your actual desires frustrated at some later time—for example, by various sorts of coercion that you might be subjected to. As an initial view, it seems that genetically engineered people can possess autonomy, in the sense currently under discussion, just like other people. Admittedly, their desires will have been shaped by such things as their genetic makeup and early upbringing as well as by the vagaries of life experience, but that is a different point—and besides, everyone's desires are shaped in such ways.

So what's the problem?

Autonomy All the Way Down?

What we must *not* do, if we wish to develop an argument based on considerations relating to autonomy, is complain that a genetically engineered embryo, or the person it will become, has been deprived of some kind of autonomy that ordinary people do not possess in any event. To expose the problem, I will introduce Galen Strawson's concept of being *self-determining* (or *self-determined*; he uses these terms interchangeably). For Strawson, self-determination in this sense is needed for true responsibility (2010, 22), and hence for what he considers the "ordinary, strong sense" of free will or free agency (ibid., 1). The assumption commonly made in the field of analytic philosophy is that we possess free will just insofar as we possess the capacity to act with moral responsibility, and Strawson does not depart from this (near) consensus.[1]

Strawson holds that none of us is self-determining in the sense that he describes. This kind of self-determination is something that we want, and routinely assume that we have, but it is an illusion. Described at a highly abstract level, Strawson's argument is that whenever somebody acts for reasons, and seems to be acting freely, the way she chooses to act is nonetheless a function of (among other things) how she *is*, mentally speaking. For someone to be responsible for *that* would require a deliberate and successful choice to be that way. To have made such a choice, though, the person would have had to exist already, and she'd have had to apply some principles of choice in the process of choosing it. These principles of choice could be preferences, values, proattitudes, ideals, and so on, in light of which she chooses how to be. But she'd have to have been responsible for *those*, also, which means choosing to have them—which would require, for a start, already existing and applying some anterior set of principles of choice. And so forth. Eventually, we reach factors beyond the agent's control that have shaped her character (Strawson 2010, 24–25).

The argument can also be formulated in a more concrete way. How we *are* is a result of our heredity and early experience, for which we cannot have been responsible (morally or otherwise). Furthermore, even if some aspects of how we are can be traced to indeterministic or random factors, we cannot be held responsible for those either. If we attempt to change ourselves, later in life, the particular ways in which we try to do so, along with our degree of success, will be determined by how we already *are*, and any further changes that we can bring about after making certain initial changes will, in turn, be determined, via the initial changes, by heredity and previous experience (plus the effects of any indeterministic influences) (ibid., 25–26).

In my view, this reasoning is unanswerable if seen as directed against a form of self-determination that includes a power of ultimate self-causation. If that kind of self-determination is needed for free will and the capacity for moral responsibility, it follows that we never possess these things. Since ultimate self-causation does not exist, neither does autonomy if we equate it with Strawson's demanding concept of self-determination. Genetically engineered individuals would not possess autonomy in any such a strong and metaphysical sense, but then again neither does anyone else. No one ever possesses autonomy if it is taken to mean some kind of *ultimate* autonomy, all the way down.

But before I continue, allow me to acknowledge that some philosophers engaged in the centuries-long debate about human free will actually do assert that we possess something like the radical capacity of self-determination that Strawson denies. In brief, there are many philosophical positions on free will, but contemporary philosophers debating the topic tend to fall into three groups: those, such as Strawson, who essentially deny the existence of free will; those who claim that we possess free will in some important sense, despite also believing such awkward things as that we do not possess a capacity for ultimate self-causation, or that we live in a deterministic universe, or that our universe is a mixture of determinism and randomness; and those who allege that we possess a more radical capacity for original acts of will, perhaps including self-determining or self-shaping acts. These might be labeled, somewhat roughly, as incompatibilist (including hard determinist), compatibilist, and libertarian positions.

I do not propose to investigate the varied and fascinating free will literature here. For those interested, a good place to start would be the current edition of Robert Kane's (2011) massive anthology, *The Oxford Handbook of Free Will*, or perhaps a less daunting collection of readings such as that compiled by Derk Pereboom (2009). The crucial point is that I know of no position in this body of literature that would assign any self-determining power, or any form of free will, moral responsibility, or the like, to ordinary people without also assigning it to genetically engineered people.

If, for example, some kind of indeterminacy within the activity of the human brain provides a basis for the exercise of libertarian free will, and with it some kind of radical self-determining, this would apply equally to the brains of genetically engineered people. While I have already expressed support for Strawson's key argument, it is difficult to see how the situation would be any different if we possessed libertarian free will: genetically engineered people would possess the same capacities as ordinary people, so the former would be no worse off. Nor does the situation appear to change if we reply to Strawson along compatibilist lines. Perhaps our actions can be free in an important sense, despite the fact that we lack a power of ultimate self-causation. But if that is true, why does it not apply equally to genetically engineered people?

Once again, I am unaware of any philosopher, bioethicist, or other thinker who has explicitly contended that ordinary people possess an ultimate power of self-causation that genetically engineered people would not. Yet some of the arguments that we will encounter in this chapter are difficult to understand unless something of the kind is being assumed (Blackford 2010, 84).

Self-Reflection

It is often argued that typical human beings possess important capacities that are less metaphysical than self-determination in Strawson's radical sense—but richer than a bare capacity to make our own decisions. At issue is the idea that we are, or can be, *self-governing* persons, and frequently there is a question of when our self-governing decisions ought to be deferred to by others, including others who intend or purport to challenge them for our own good. For John Christman (2009, 162), for example, a central issue is when we can rightly consider an agent to be someone whose capacities and viewpoint "should matter as the sources of valid claims in collective decisions and toward whom paternalistic intervention would be disrespectful."

This idea of autonomy as self-governance, or an especially legitimate form of self-governance that merits deference from others, has great philosophical and practical significance. In particular, it is a fundamental concept in modern medical ethics and related fields. Medical practice and health policy are frequently supposed to be constrained in substantial and important ways by ideas of autonomy. Even people who are incompatibilists about free will are likely to rely on such a concept. Thus, a philosopher might deny the existence of free will, yet still protest if a doctor treats her problems in a way that is against her wishes or which the doctor refuses to explain.

One important line of thought connects autonomy and hence self-governance with the ability to reflect on our own actions, not merely seeking to achieve our immediate desires, but instead going further and considering whether we really *want* those desires. This can extend so far as reflecting on whether our current dispositions of character are what we want, or whether we wish to alter them in some way.

There is a formidable literature relating to this, much of which builds on Harry Frankfurt's analysis of free will and related ideas. This is most readily available in Frankfurt's essay collection *The Importance of What We Care About*. Frankfurt (1989, 16–20) understands our freedom as the ability to order our wills in accordance with what he calls second-order volitions—that is, with second-order desires to have certain kinds of first-order desires (and not just to *experience* them, but rather to have them so that they become the agent's will). By contrast with such an exercise of freedom, he describes situations where we could have emotional responses that are (in a sense) external to us, such as those induced by hypnosis or the use of a drug—in which case they do not arise as responses to a perceived experience—or those that take the form of, say, an out-of-character outburst of temper that feels alien to the speaker (ibid., 62–63).

Thus, Gerald Dworkin understands autonomy as an individual's capacity to reflect on her first-order preferences, wishes, desires, and so forth, and accept these, or else attempt to shape them in light of higher-order ones. He contends that this can give someone's life meaning and coherence, as she takes responsibility for the kind of person she is (Dworkin 1988, 108–109). Michael E. Bratman's (2007) recent collection of essays on the subject of autonomy engages closely with Frankfurt's views, stressing not only the structured and hierarchical character of our desires and values but also the key role of *planning* in human activity.

Glover's account of human motivation is slightly different from this, but (I think) essentially compatible with it. He suggests that it is plausible that a person's strongest desire *at the time* always prevails at that time. Yet, he maintains, there is something inadequate about identifying a person with her pattern of desires at a particular moment, since we all have longer-term plans that form a basis for second-order desires about what desires will prevail, and therefore enable us to restrain some of our immediate impulses. In identifying with a higher-order desire, we think that acting on it will be in accordance with what we generally most care about, or our picture of the person we want to be (Glover 1988, 125–126, 150–151).

This discussion by no means exhausts the rich lode of work by philosophers who attempt to understand autonomy in ways that involve ideas of reflection, integration, and so on, without self-determination or self-creation all the way down. To take just one more example, Diana Tietjens Meyers's work is pervaded by the idea that autonomy is possible even though human beings are shaped by socialization within their respective cultures. Meyers (2004, 257–273) provides a concise description of autonomy, seen in terms of self-chosen goals along with congruent decisions and actions, in her essay "Gendered Work and Individual Autonomy."

Nor have I yet acknowledged the radical challenges to ideas of self-governance that can be found in much continental and postmodern philosophy. As Christman (2009, 51) expresses it, a prominent theme in postmodern thought is that "the self is constructed out of the power dynamics of present and past social structures." From this viewpoint, there is a continuing interplay between conceptual categories and institutional structures, as they tend to shape each other over time. We are, ourselves, deeply, unconsciously, and always-already shaped by language, institutions, social settings, and the ubiquitous demands and judgments of others. On the face of it, this leaves little, if any, room for the existence of rational, self-reflective, relatively stable, self-governing persons.

Christman seeks to offer a philosophical account of autonomy, while largely accepting or at least entertaining this picture. How far he succeeds may be open to debate, but it should be noted that this postmodern doubt about individual self-governance applies equally to the lives of ordinary people and any genetically engineered people who might come into existence. If we share the doubt, we should doubt whether *anyone* is truly self-governing, whether or not her genetic potential is partly a product of human technological intervention.

For the sake of this book, I take it that concepts of self-governance are compatible with whatever truth might be discernible in postmodern, skeptical accounts of the self and the ways in which it is constituted. On that assumption, competent adult human beings (and children beyond a certain stage of cognitive development) are capable of making their own decisions, provided they are not subjected to coercion or other activities that undermine their will. They also possess something more: a rather rich capacity for self-reflection and self-direction. Moreover, we place a high value on these capacities; we want to retain them for ourselves, and we want others to develop them.

None of this assists critics of genetic engineering, and I am unaware of any critic who has specifically argued that genetically engineered children would fail to develop capacities for self-reflection and the like. For example, there is no reason to imagine that such a person would (any more than the rest of us) develop special cravings that her second-order volitions could not overcome, or irresistible impulses that she would yield to automatically. While it is conceivable that some *specific kinds* of genetic interventions might damage the capacity for reflection—think, say, of an intervention to reduce the individual's potential for intelligence to that of a monkey or a dog—a wide range of genetic interventions would have no such effect. For instance, somebody whose genome was tweaked at the embryonic stage to avoid a propensity for certain diseases would not thereby lose the normal human capacity for self-reflection. The same applies to someone with the potential to develop high intelligence, great physical strength, or extraordinary disease-resistance or longevity.

The situation, then, is that *neither* an ordinary nor a genetically engineered person will be able to step out of the entire effects of her history, molding by social forces, or total set of current beliefs and desires. Nevertheless, *both* will (probably) be capable of self-reflection and self-direction. This includes at least some ability to engage in rational appraisal of ourselves as well as our actions, and make efforts to shape our own dispositions. As Stephen Holland (2003, 181)—an opponent of genetic engineering—acknowledges, an individual will not be less capable of this kind of rational appraisal if she has been genetically altered than if she has not.

Self-Creation and Independence

Elsewhere in his work, Glover (2006, 69–72) looks at a power of self-creation that is, as he notes, never total, since it never amounts to ultimate self-causation. Self-creation, in Glover's sense, can be thought of as possessing the scope to shape our own identities, with relatively open future possibilities—something that might actually be increased by having greater talents or abilities. Yet Glover suggests that *two* important values are at stake in the debate about parental intervention in the genes of children. In addition to our scope to shape ourselves, there is what he calls "independence," the idea of our nature's not being merely the product of decisions by other people. Glover points out that these values—self-creation and independence—can actually come into mutual conflict if an individual's genes are shaped by her parents, but in a way that increases her capacity to create her own future. Independence can also conflict with other values; for example, others might influence someone in a way that gives them a capacity for richer experiences of the good things of life.

Glover suggests that too much genetic intervention might make us feel like the puppets of our parents, although he acknowledges that there are other values at stake, and proposes that we might accept some loss of independence, in the relevant sense, in exchange for a richer life or greater power to shape ourselves.

Perhaps one of the issues here is the distinction between determining and merely influencing. In some extreme cases, where parents could totally control how their children turned out, gaining power over every genetic and environmental influence to produce a detailed, preconceived outcome, we might object. We might maintain the objection even if the effect of the influences were an expansion of abilities. Even now, we might be concerned at a child being given intensive lessons in violin playing or classical languages at the age of three (although it is far less clear that this is something we'd wish to prohibit).

But what is the objection to violin lessons for infants based on? Perhaps it is a sense that spontaneous play is important to childhood and should not be crowded out by other things, as it was in Mill's famously precocious infancy. Perhaps it is a sense that too much childhood regimentation is harmful to children's flourishing. If something like that is the problem, it should not color our attitude toward genetic enhancements that expand abilities, especially if we are talking about enhancements that are general-purpose goods. It is, of course, *possible* that parents who care deeply about their children's developing abilities will subject their children to damaging kinds of regimentation, yet it by no means follows. After all, we do not assume that someone who takes the trouble to provide a child with good nutrition is also acting in ways that take the play and spontaneity out the child's life. Why should we think differently about someone who has augmented her child's genetic potential?

Some critics of genetic intervention may be most worried by a loss of independence, as defined by Glover. It is difficult to be certain of this, since the critics seldom explain in any detail just *how* genetic interventions will threaten autonomy. Assuming that this kind of independence is the main value at stake, I am not convinced that it is of overriding or fundamental importance. Michael J. Sandel (2007, 81–83) suggests that we must be able to think of our origins as having a beginning that is beyond another's disposal if we are to think of ourselves as free, but he offers no empirical support for this claim, and it is not self-evident. It is true, no doubt, that we would all fear falling into the hands of a devil/neurologist who could reshape our fundamental desires (Frankfurt 1989,

52–54), were such a thing to become a serious prospect. That, however, is at least partly because future interferences of such a kind would destroy our present personalities and frustrate all our current hopes about how we want to shape ourselves and pursue our projects.

We do not regard the *past* actions of parents in the same way, even though, in the normal course of events, they have done much to shape our current personalities. We can recognize the ways in which we have been molded by parental choices without reaching a conclusion (such as Sandel's reasoning would seem to prod us toward) that we are not free.

It is unavoidable that parents shape their children's characteristics, employing such means as nutrition, education, discipline, and socialization into a culture. Admittedly, some parental actions may have only trivial effects, and some may miscarry, but others may have deep and far-reaching effects on the beliefs, capacities, and dispositions of a developing child. While all of these will also be affected by genetic determinants, the latter may actually have *less* impact on the child's sense of identity, including her understanding of the world and her place in it, what she values and loves (and what she despises), what goals she has in life, what skills she develops to pursue them, and how she is disposed to act in the range of situations that confront her.

To be fair, genetics might also have *more* impact on these in some cases; how somebody turns out might be affected in numerous ways by her genetic potential. But this should be kept in perspective: people with widely different genetic potentials may end up having similar values, worldviews, conceptions of the good, and so on, if they have all been immersed as children in the same culture.

In short, it would be a serious error to see the genetic determinants of our characteristics as more fundamental than environmental ones. The human genotype directs our development down a certain path that involves, for example, a basic bodily shape and species-typical kind of neurological development. Yet the variations among human beings that matter most to individual identity and style are largely cultural, or otherwise environmental. At the same time, even the fulfillment of someone's genetic potential for such gross characteristics as height, weight, and strength depends on the environment in which the genes are expressed. In all these cases, how we *are* does not depend merely on genes or solely on environment but rather on the interaction of both.

The personalities and abilities that individual people end up with as adults are currently influenced, in every case, by a variety of past circumstances that were beyond their control, particularly the actions of parents. If parents did gain some level of ability that they currently lack—that is, some ability to control their children's genetic potentials in addition to the environment in which the potentials are expressed—this would not necessarily be harmful to children, even apart from the prospect that parents might use their newfound control in ways that are actually expansive of potentials.

Perhaps one could speculate about the possibility of psychological harm, combining this with precautionary thinking to reach a conclusion that prohibition is required until such a time as more is known. But this seems too cautious, especially if the prohibition is aimed at a broad range of interventions, including those that involve general-purpose enhancements. Consider a specific example: suppose that Arnold, now at the age of nineteen, discovers that his parents *not only* provided him with sound, protein-rich nutrition and plenty of opportunity for exercise but *also* tweaked his genome to give him a greater potential to develop strong muscles. He has turned out to be a powerfully built young man. Why, exactly, should he feel burdened in any way because the interventions made by his parents were genetic *as well as* environmental?

One answer might rely on Glover's point that excessive intervention by our parents could make us feel like puppets, yet this is surely a matter of circumstance and degree, and (to some extent) speculation. It is possible, let us suppose, that the sheer comprehensiveness of parental control could make a child feel weirdly like the parent's product, but that is not the case with Arnold. His parents' actions leave open what worldview he might come to accept, what creative activities and other pleasures he might take up, what attitudes he might bring to such conflicting priorities as work and family, and what temperament he might have. There is no comprehensive control here. I see no good reason for Arnold to feel like a puppet.

The situation in fact might be different for a child who has been comprehensively designed for such things as the trivialities of physical appearance—I am thinking not just of health, body and facial symmetry, and whatever else might be the broad cross-cultural foundations of beauty but also of such things as whether her eyes should be green or blue, the exact length of her nose, and so forth. Perhaps there are extreme cases where a child could understandably feel like her parents' puppet or dress-up doll.

Then there might be attempts to shape a child's psychological identity, such as her sexuality, her degree of religiosity, or her precise traits of character. Such attempts are already made by environmental means, and they are no doubt successful in many cases, as with the large number of devout Christians, Muslims, Hindus, and others who manage to rear their children to be equally devout in the parents' faith. Again, large numbers of parents manage to inculcate in their children similar attitudes to their own when it comes to sex and sexuality—not to mention sports, art, or business practices, and many other things. The law does not interfere with any of this, so it is not clear why we should feel more concern where the influence is exercised partly by means of genetic interventions.

The Open Future Argument

As Glover brings out in his exploration, the idea of self-creation is closely related to that of an open future for the child. The notion here is to enable the child not only "to perform his part well in life towards others and himself," as Mill (1974, 176) puts it, but beyond this, to be well placed to choose from a varied array of life plans. The concept was formulated by Feinberg (1992, 76–97) in an article first published in 1980: "The Child's Right to an Open Future." Feinberg's piece discussed the US Supreme Court's decision in *Wisconsin v. Yoder* to allow Amish families to withdraw their children early from Wisconsin's education system, which required compulsory school attendance until the age of sixteen.[2]

The idea of an open future is, of course, difficult to reconcile with detrimental genetic modifications, but what if the modifications are beneficial (as with a potential for unusually great longevity)? What if they are more or less neutral in their benefit or detriment, such as the choice of a genetic variation for green eyes? As discussed above, the latter choice might seem problematic to the degree that the child could come to feel that she had been treated like a toy, but that is a different issue. Leaving that to one side, the possession of green eyes, or eyes of any other color, is quite compatible with possessing such capacities as those for rational judgment, self-reflection, autonomous choice, and so on, and with acquiring an indefinitely wide range of skills and capacities.

Moreover, genetic engineering could actually enlarge children's futures to the extent that it could enhance their capacities to think rationally and acquire a range of useful skills as well as broadening and deepening their understanding of the world and their place in it. These would seem to be natural effects of interventions designed to increase cognitive capacities, health, or longevity—and perhaps even such things as perceptual abilities and physical strength, which underlie the development of many skills. Thus, Nick Bostrom (2005, 212) writes, I think persuasively, "Being healthy, smarter, having a wide range of talents, or possessing greater powers of self-control are blessings that tend to open more life paths than they block."

There is, certainly, a difficulty in talking about someone's future being more or less "open," since the capacities and skills required for some
choices will hinder or preclude other choices. Still, this nicety does not normally deter us from praising parents who provide children with exercise,
education, and good nutrition, partly because such actions are thought to
open up options for children. Nor should it deter us from praising parents
who provide their children with benefits to their genetic potential.

The argument, however, might be run differently. One concern is that
any parent who has gone to the trouble of attempting to control a child's
genetic makeup is likely to be overly controlling of the child's emerging
personality and life prospects. Another is that the child herself might feel
constrained by her knowledge of having been genetically modified, or
in the case of reproductive cloning, having the same genes as an earlier
genetic twin. In the context of reproductive cloning, Dan Brock (2003,
639–640) mentions the possibility of psychological pressures to match
the achievements of an earlier twin, if these were exemplary. Similarly,
Kass (2001, 34) insists that people will always be comparing the clone's
doings in life with those of the person who was cloned.

Are these good reasons for a hostile regulatory attitude? Brock (2003,
637–639) suggests that a child retains an open future as long as she really
is free to choose, even if someone misleads her, intentionally or otherwise,
to believe that her future has been closed and determined. This seems to
be an inadequate response. Surely a person who has internalized an idea
of herself as having a closed future might thereby become psychologically
incapable of choosing, even if she has the necessary array of skills to pursue a variety of life plans. This appears to be a real concern, and it does
not seem helpful to insist that a psychologically burdened person "really"
has an open future, unbeknownst to her and beyond her practical access.
Indeed, depression (when it occurs in reaction to some significant loss)
seems to work much like this.

Nevertheless, children can and often do rebel against parental expectations, particularly against being expected to follow a preexisting life
pattern. Matteo Mameli suggests that this would continue to be the case
even with cloned or genetically engineered children. In any event, he says,
we could teach children from an early age that having the same genes
as someone else does not predestine an individual to leading a similar
life (Mameli 2007, 91–92). All things considered, perhaps we ought to
remain disheartened by the prospect that some children could suffer from
the actions of foolish or obsessive parents. But at the same time, it would
be highly illiberal to enact legal prohibitions—prohibitions applicable to
all parents or potential parents contemplating the use of enhancement

technologies, and a broad class of actions that they might take—merely because *some* parents might be foolish or obsessive.

Furthermore, the open-future argument seems especially misguided when related to interventions that could boost a child's potential to develop abilities and skills. A choice to increase a child's genetic potential for musicality, for example, does not necessarily go hand in hand with any desire for her to take up a career as, say, a singer or violist with an orchestra. Such an enhancement could be useful for a vast range of life plans. Successful attempts to increase a child's physical strength, mathematical ability, or resistance to disease would be even more likely to assist her in a multitude of life plans, and need not signal any attempt to be unduly controlling.

Detrimental Interventions

Up to now I have been exploring the alleged threats to autonomy from genetic intervention itself, rather than the possibility of truly detrimental modifications. It is conceivable that somebody could use genetic engineering to reduce her child's cognitive potential with the intention or effect of damaging the developing individual's capacities to exercise such capacities as those for autonomous choice and self-reflection.

Genetic engineering might be used by some parents with bizarre values to create children with reduced problem-solving abilities or an impeded capacity for communication. Less dramatically, but perhaps more plausibly, one could easily imagine that some parents might attempt to boost such characteristics as docility, deference to authority, or childhood credulity in order to facilitate the shaping of a child's temperament. Other choices, such as to give the child a minor or major physical disability (such as a speech impediment or deafness), might have similar effects more indirectly, such as by creating more dependency on others.

Note, though, that the problem here is the detrimental nature of some specific interventions that we can imagine; it is not a problem for genetic intervention as such. Moreover, we should not be quick to label possible interventions as detrimental. For example, an intervention that predisposed a developing individual to more friendly and cooperative behavior does not seem morally analogous to an intervention to predispose someone to credulity, docility, or undue deference. If we set aside *how* someone's personality took shape, having a friendly, cooperative disposition is not, in itself, the sort of thing that undermines or hinders the capacity for rational reflection. We normally recognize this sort of disposition as compatible with the exercise of autonomy (Walters and Palmer 1997, 127).

Habermas and Somerville on Autonomy

A Threat to Liberal Society?

We are now in a position to examine the approach taken by Habermas, who argues that the genetic engineering of human embryos would damage the autonomy of individuals in a way that would, in turn, threaten values important to modern liberal societies. He expresses the point in various ways, such as when he claims that an irreversible decision affecting another person's "organic disposition" will restrict "the fundamental symmetry that exists among free and equal persons" in modern societies. Or, he says, it could change our self-understanding so as to undermine "the inalienable normative foundations of social integration." In particular, it could undermine our self-understanding as persons leading our own lives and showing each other equal respect. Or we will be unable to come to a self-understanding as being "the undivided authors of our lives" and to regard all others "as persons of equal birth." Yet again, Habermas suggests that genetically engineered individuals may not be able to "regard themselves as sole authors of their own life history" as well "as unconditionally equal-born persons in relation to previous generations."[3]

At the heart of all this is the claim that someone whose genome has been, in part, designed can never exchange roles with the designer, and so has a permanent social dependence that is incompatible with the reciprocal, symmetrical, and egalitarian relations that supposedly underpin morality and law in modern societies. As already mentioned, however, no one is ever the undivided author of her life. Liberal democracies cannot rely on such a flimsy idea as that. This is also fatal to an argument developed by Linda Barclay (2003, 230–231): the widespread use of genetic engineering might encourage the idea that genes can profoundly affect or even determine how people turn out, leaving them with little sense that it is their own agency that is primarily responsible for their significant desires and values.

Why should people have any "sense" of something like *that*? Since none of us are self-creators all the way down, it is quite misleading to describe ourselves as "primarily responsible" for our own desires and values. Surprisingly, this point has passed largely unremarked in the relevant bioethical debates, although one philosopher who does draw attention to it is C.A.J. Coady. He observes: "No one can regard themselves as the sole authors of their own life history, there is too much contingency and inevitable dependence on others for that to be plausible" (Coady 2009, 174).

Our significant desires and values are, admittedly, not solely genetic in origin, but they are the product of a combination of sources that lie outside of ourselves (biological, familial, and social). We have some capacity to shape our desires and values, but not to step outside all of them at once. (And even if we have, at least in part, created ourselves through acts of libertarian free will, this ability would not be affected one way or the other by whether or not our genetic potentials were chosen by our parents.)

Might it be said, in reply, that though all this is true, it is important for us to retain the comforting illusion that we possess complete autonomy? Might we even have a duty to promulgate the idea as a noble lie? Based on this approach, the genetic engineering of human beings would be wrong because it would tend to dispel the illusion and hinder attempts to disseminate the lie. I have not seen the argument put that bluntly and candidly, but something like it might be offered in Habermas's defense.

The offensive paternalism of such an approach is not conclusive against it; perhaps it is sometimes justifiable on utilitarian grounds (for example) to engage in offensive paternalism. But it is doubtful, to say the least, that such an insulting approach to public policy is necessary. Harris has remarked that we have managed to get used to the idea that we have evolved from apes and are not at the center of the cosmos. He adds that it is not obvious that we need to regard ourselves as entirely the products of nature and in no way products of human design (Harris 2007, 154–155). Perhaps he exaggerates human adaptability, since rearguard attempts are made by religious fundamentalists in many countries to resist the idea that human beings evolved from earlier primates. Nonetheless, the force of his comment is apparent: it is possible for human beings to cope with truths that undermine traditional images of humanity's place in the natural world.

Additionally, Habermas is wrong to suggest that the foundations of modern society are incompatible with various kinds of asymmetry and inequality, such as those between parents and children. As Elizabeth Fenton (2006, 39–40) observes, there is an inherent inequality between parents and children in any event, given that children are dependent on parents for their existence; it is, moreover, neither avoidable nor undesirable that parents attempt to shape how their children will turn out. Indeed, it remains common, if no longer so typical, for parents to socialize their children into a more or less identifiable body of beliefs, traditions, and cultural values. Although children can be demanding, it is far from usual for *children* to socialize *parents*, or parents and children to exercise an equal, mutual influence.

This is not to deny the crucial point that children may show certain inclinations and talents from an early stage. As Feinberg points out, they exhibit a sort of rudimentary character even from birth. Wise parents will respect natural biases in their children's development that arise from heredity and early environment, shaping them with the grain of their developing abilities and proclivities, rather than against it. Thus the process of socialization is dialectical to an extent: the child's character develops in response to the initiatives of the parent at each stage, yet these, in turn, are chosen partly in response to the development of the child's character in one direction or another. The relationship is not simply one in which the child is passively molded but instead is more subtly cooperative (Feinberg 1992, 95–97).

Feinberg's description appears somewhat idealistic, but let us set that aside. He no doubt is correct that there are important ways in which the socialization of a child may *respond* to her developing character as opposed to creating it in toto. But this also applies to genetically engineered children! Imagine that Belinda's developing abilities and proclivities have *not only* been influenced by decisions about her early environment (such as a decision by her mother, Abigail, not to drink heavily during pregnancy, or to enrich Belinda's home environment by providing toys and plenty of interaction with adults). Assume that Belinda's abilities and proclivities have *also* been influenced by genetic modifications to enhance her potential for intelligence, longevity, and physical strength. Nonetheless, once she begins to develop a character of her own, a wise parent will follow Feinberg's advice and socialize her in a way that goes with the grain as far as possible.

In any event, Habermas exaggerates the significance of equality to modern social life, with its many political and economic hierarchies and its uneven distribution of wealth. Even if these are objectionable, they do not render modern society unworkable. A weaker and more accurate claim that Habermas might have made is that liberal societies are committed to eliminating certain specific kinds of arbitrary subordination (e.g., on the grounds of race, sex, or sexual preference) and ensuring that new ones do not arise. Yet *this* claim would not assist his argument, since the kind of inequality that he is discussing—the asymmetry between parents and children—creates no obvious likelihood that genetically engineered individuals will become a new subordinated class. If anything, the more plausible fear is that genetically engineered individuals might become a dominant caste, subordinating the rest of us. Indeed, Habermas (2003, 81) basically concedes the point when he acknowledges that it is implausible to claim that the "objectifying" attitude of the parental designer to a designer child will continue after the birth.

Once this is conceded, what is left of the argument? Habermas states that the issue is not one of discrimination that a genetically engineered person would suffer, any loss of basic goods or life options, or vulnerability to being forced into particular practices. Rather, it is that she might feel a lack of a mental precondition for meeting the expectation of taking sole responsibility for her life (ibid., 81–82). But this simply returns us to the fantasy that anyone ever takes, or *could* take, such "sole responsibility," or that the state should legislate on the basis of such a philosophically controversial idea.

Habermas also makes a strong claim about the proper limits of parental authority to initiate any genetic intervention. Could this be used to ground a claim that genetically engineered people will experience a special kind of subordination that is incompatible with liberal society? He asserts that genetic intervention could be justified only if the informed consent of the embryo could be imputed, which (he believes) could only be in extreme cases, such as to save the individual from a life of extreme suffering (ibid., 43–44, 63, 91–92). If this criterion were adopted as a legal standard, it would be highly restrictive, even of genetic modifications aimed at ensuring good health. Still, Habermas misconstrues the rational basis for the doctrine of informed consent. As competent adults, we rationally fear having intrusive medical interventions imposed on us without our understanding or against our will—even if, in reality, they are for our own good. Children who are growing in maturity and understanding can have similar fears. An *embryo*, however, cannot fear anything like this.

If I am charged with the responsibility for giving an embryo (or the person it will become) the best start in life, the more relevant questions are the following: What is in the interests of the child who will come to be? What sorts of actions would a reasonable child thank me for later on as she comes to understand why I acted as I did? At the same time, what forms of political coercion do I have good reason to fear in this kind of circumstance?

As Habermas's own discussion brings out, the interventions that I might be able to make could be highly advantageous to the person to be. They might well give potentialities that a reasonable child would thank me for later in life. These might include a longer life span or better immune system, such capacities as increased musical talent, or highly generalized capacities like enhanced intelligence, strength, and memory. One of the things that reasonable parents or others with similar responsibilities might wish to be protected from is the possibility that some meddler—perhaps with political power and supported by an electoral majority—might find ways to constrain their ability to grant gifts like these.

The Essence of Humanness

In chapter 1, I referred to Somerville's apocalyptic claim that genetic engineering would destroy "the essence of the humanness of us all." Her concern is actually similar to that of Habermas, and it is vulnerable to much the same criticisms. First, she claims that we need to have a sense that we can remake and actualize ourselves—which may be true, but I have argued throughout this chapter that it is of limited relevance to policy debates about genetic engineering. She further alleges that our ability to take part in human interaction is dependent on being free from others' interference in our "intrinsic being," as she puts it. Moreover, she asserts, we are undermining democracy and liberalism if we impose our will on future generations, and even worse, we are all complicit in these evils if we do not take steps to prohibit genetic engineering (Somerville 2007, 145–146).

These claims are vague, overly rhetorical, and (to the extent that their meaning can be ascertained) false. For a start, it is doubtless true that many people value the ability to remake their own personalities. But at the same time, we are all limited in how far we can do this; the most we can do is make some limited changes that are possible "from here," given the values and other principles of choice that we already have. There is no reason to believe, though, that genetically engineered people will be less capable than others of this kind of limited self-alteration. Since our genetic potential (which includes the potential for athletic ability, intelligence, longevity, and so on) provides one set of constraints on how far we can change ourselves in various directions, it is entirely possible that some genetically engineered people will actually possess greater than usual opportunities for self-creation of the relevant kind.

Furthermore, it should be clear by now that there is no reason at all to believe that we need to be free from interference in our "intrinsic being" in order to take part in human interaction. For example, many people are taught traditional values, supposedly virtuous dispositions, and religious beliefs that all seem more essential to their intrinsic being as individuals than such things as their genetic potential for strength or intelligence. Some of these teachings may be unfortunate in many ways—they may involve false ideas about the universe, dubious virtues such as chastity and piety, or pernicious systems of values—but except in extreme cases, it cannot be said that people who have thoroughly internalized them are thereby rendered unfit for human interaction.

If this is thought to be quibbling, and if we take the expression *intrinsic being* as roughly equivalent to genetic potential, then by all means let us consider the following claim: "People whose genetic potential has been interfered with by others are unable to take part in human interaction." But there is no reason to believe such a thing. We might just as well believe a claim like "people who have been interfered with (environmentally) in the development of their genetic potential are unable to take part in human interaction." While it may be difficult to specify what actually does equip us for human interaction, it is more likely to be certain of our developed phenotypic characteristics—such things as imagination, empathy, and communicative capacities. Yet there is no reason to assume that genetic engineering will damage these.

As for its being contrary to democracy to impose our wills on future generations, this suggests a bizarre concept of democracy in which the unborn generations should be given a vote about present-day decisions. It is, of course, true that many of our decisions, individually, cumulatively, or collectively, will affect future generations in important ways. For example, our decisions can affect the number of people there will be in the future, what kind of physical environments they will inherit, what traditions of art, science, and scholarship will be handed to them, and so forth. The eradication of a disease such as smallpox affects future generations, but we assume that they will be thankful to live in a world without this blight. Every action taken by anyone has some small impact on the physical, social, political, and moral environment of the future, with cascading effects that will continue indefinitely and can never be predicted with any great accuracy.

We cannot avoid making decisions that affect future people, while being conscious that they can never participate in an intertemporal electoral process to influence the outcomes of our various deliberations. After all, future people don't even exist as yet, and *which* people will come into existence is highly sensitive to our actions. We cannot consult their actual wishes. The most we can do is recognize that there will (probably) be some people in the future, and we can affect whether they will be people with happy, flourishing lives or the opposite. Those of us who are alive now may all agree to place value on the welfare of future generations, but that shapes *all* our decisions. It is not a reason to prefer that the genetic potentials they inherit be left to nature, with no contribution from deliberate genetic engineering decisions.

A More Charitable Reconstruction?

Can the arguments developed by Habermas (and echoed by Somerville) be reconstructed more charitably than I have succeeded in doing so far? To return to this chapter's fundamentals, Habermas needs to distinguish the use of enhancement technologies from the many choices that parents make about upbringing with the aim of showing that the former involves a special threat to autonomy. One way might begin with the claim that psychological and other outcomes produced by upbringing can be challenged—they allow for a revisionary self-understanding that may deviate from or even oppose the will of the parents—while changes produced by genetic intervention cannot. There are numerous problems with this, however.

To begin with, environmental effects on psychological development are often not reversible, while many genetic effects on it *are* (Mameli 2007, 89). As for nonpsychological traits, vaccination triggers an immune reaction that permanently alters the operation of the immune system, while many other environmental interventions have crucial and irrevocable effects on a child's body shape, muscular development, neurological development and connectivity, and cognitive and emotional characteristics (Buchanan et al. 2000, 160; Harris 2007, 139). It is not a simple matter to reverse any of these (Glover 1984, 53).

Moreover, we might doubt whether reversibility in itself is so desirable. Glover has asked whether it would be a good thing if efforts to inculcate kindness and generosity in children could be reversed. It seems not (ibid.). There may be many outcomes from *environmental* interventions that it would actually be undesirable to reverse, even if it were possible. These outcomes include good skeletal and muscular development, immunity to certain diseases (from vaccination), the development of cognitive capacities and skills, and socialization into basic virtues, such as a degree of kindness, generosity, and cooperativeness. Insofar as these are difficult to reverse (except by brutal means such as inflicting brain damage), that seems to be a *good* thing. All else being equal, if making genetic as well as environmental interventions could assist these outcomes, it also would appear to be a good thing. Habermas provides no persuasive argument to the contrary, and no reason why any difficulties in reversing such interventions should be seen as a special threat to the autonomy of individuals or functioning of liberal societies.

Perhaps Habermas could attempt to make a different claim, as Barclay does, about the ideal way to educate and socialize children. According to

Barclay, we can and ought to give children a moral education by offering them reasons why some things are right and others are wrong. Children might later reject these, contends Barclay (2003, 232–233), so it is not an exercise in control but instead something more like an exercise in rational persuasion. This sort of reasons-based approach to upbringing can then be contrasted with more coercive attempts to mold a child's personality via genetic engineering.

Harris (2011, especially 102–105, 110–111) appears to raise similar problems in a recent critique of the idea of "moral enhancement," in which he expresses concern about high-minded attempts to eliminate our freedom to act badly. As I read his argument, however, his main emphasis seems to be on the difficulty of producing truly effective modifications to whatever might be the psychological underpinnings of human morality, combined with what he sees as the dangers of postponing efforts at *cognitive* enhancement. If I understand him correctly, Harris is most worried about foregoing, or postponing, efforts that could lead to highly beneficial advances in scientific and intellectual understanding.[4]

In any event, Barclay makes a straightforward enough case against moral enhancement as a threat to autonomy. Should we accept the distinction that she relies on? For the present purposes, the distinction must not only be clear and intellectually defensible. It must point beyond that to a threat to the autonomy of individuals and social functioning. Here I am skeptical.

Barclay is making a narrow assertion about a problem with, say, using genetics to "program" a child's behavior. Compared to a child with the reasons-based upbringing that Barclay advocates, the child whose thinking is genetically programmed in this way will be less able to reject her parents' ideas of virtue and moral goodness, and thus is more deeply subordinated. I am skeptical about how all this is supposed to work, but note that the argument is confined in its application. It is, as far as I can see, simply irrelevant to many genetic interventions that someone like Habermas would disapprove of. For example, it has no application to genetic engineering aimed at preventing disease (perhaps in circumstances involving an expectation less horrible than a future life of pain), enhancing athletic or cognitive potential, and so on. Admittedly, these interventions would have *some* indirect effects on the development of a child's personality and characteristic behavior—but they would not be predictable, and this looks no different in principle from the indirect effects on personality that flow from vaccination, training in sports or self-defense, teaching a child to play chess, or providing her with protein-rich food.

Even in its narrow area of application, Barclay's observation is highly contentious. Consider the inculcation of moral virtue. We really have no choice but to send messages to children about values, preferred behaviors, and acceptable dispositions of character, beginning when the children are far too young to understand any justificatory arguments. We may influence them through such methods as punishments and exposure to the behavior of good role models (Buchanan 2011b, 110–111). We *want* to instill in children some core, persistent values by which they can later judge and control their immediate impulses. We do, of course, wish to teach children to think critically for themselves, and there is a balance to be struck here (Law 2007, 34–36). At the same time, it is not at all obvious that being born with a tendency toward, say, altruism precludes learning to think critically.

Perhaps it is not too controversial to acknowledge that reliance on harsh punishments, with little or no accompanying explanations, is an ill-advised approach to child rearing. Samuel Oliner and Pearl Oliner adduce evidence that we should draw the attention of children to the consequences of their behavior, referring to other people's feelings, thoughts, and welfare. They conclude that this is a sound approach to producing compassionate, altruistic adults, able to adopt the perspectives of other human beings (Oliner and Oliner 1988, 178–183, 249–251). Note, though, how it appeals to the child's original capacity for empathy with others rather than relying on anything like an exercise of Kantian reasoning.

In the end, Barclay's distinction is circumscribed in the kinds of interventions it would condemn, and far too contentious to provide an adequate basis for a public policy that forbids even them. Perhaps we should, indeed, adopt the Oliners' suggested approach to child rearing, giving children clear, simple explanations as to why they should behave in some ways and not others. Yet it is unlikely that these explanations can refer to reason alone. They will depend to an extent on common tendencies in human psychology, such as the capacity for empathy. But in that case, why not improve that capacity, at least in any instances where we can identify a low genetic potential for it?

A Residue of Concern

Is all this analysis totally conclusive? Perhaps a residue of concern remains. Habermas and his allies in this debate might have a final response: perhaps what matters is not the *truth* of any of this. What matters, rather, is how it would seem to children (perhaps even mistakenly). For all that has been said, some children *might* end up devaluing themselves, albeit irrationally, if they knew that they had undergone genetic engineering when they were still embryos. We cannot rule out all possible situations where parents and society in general might fail to disabuse them.

Furthermore, I have already acknowledged concerns about interventions aimed squarely at influencing personality traits, or otherwise conveying an impression that the parents have treated their child as a kind of puppet or plaything. This point should not be pushed too far since it is speculative. Against such speculation, Pence (2000, 70–71) cites a body of research that casts doubt on what is psychologically harmful to children and our ability to predict this, as opposed to guessing and speculating. In a related context, he raises another issue that may be more telling, and one that surely occurs to many children, especially those who were born before highly effective means of contraception became so widespread in Western societies. Children can question, "Was I planned and wanted, or was I an 'accident'?" But not many people seem to be haunted by this question, or whatever answer they have been given as individuals (Pence 2004, 102–103).

Thus I sympathize with Buchanan's (2011b, 5) dismissal of what he calls "outmoded, armchair psychology." Genetically engineered people can be just as free and autonomous as ordinary people, and we have no evidence that they will be psychologically incapable of regarding themselves as such. They will, moreover, constantly have the experience of making choices that feel free in the same way as the choices of ordinary people feel free from the inside. All in all, we have good reason to think that genetically engineered people will take knowledge of their origins in stride.

Nonetheless, in the absence of specifically relevant empirical data one way or the other, the concern will not go away entirely. I suggest, accordingly, that we exercise caution before going down the path of modifying embryos for the purpose of producing direct alterations to children's personalities. To be clear about this, I believe that I've shown the arguments to be rather shaky. Nonetheless, until we have more data about the psychological effects of reasonably comparable interventions, there is room for sincere and intelligent concern about *this* specific type of intervention—although not necessarily to an extent that justifies legal prohibition.

The use of PGD, however, seems to be a different matter. A child may grow up knowing that she was chosen, as an embryo, partly for her potential to develop a particular kind of personality. Yet, her genetic potential has not been *tampered with*; it is simply the outcome of her being the individual that she is. There is no basis here for her to think of herself as a puppet.

Finally, even if it turned out that some genetically engineered children actually had *enhanced* capacities for such things as rational consideration and reflection, they might not be seen that way by others—perhaps because of persistent ideas about genetic determinism—and this mistake on the part of others could end up being disadvantageous to them. We are back in the territory of psychological threats to new kinds of people from social misunderstandings.

In discussing reproductive cloning, Agar argues that genetic determinism seems to persist in the face of debunking. Perhaps, he suggests, this is because "cool," "creepy," or alluring ideas are more persistent than merely true ones, and ideas about clones could be like this: there could be widespread irrational fears, say, that a person's clones are identical to them or have the same thoughts. It would be a terrible outcome for the children concerned if they were marginalized and stigmatized as a result of such irrational responses (Agar 2002, 161–168). If Agar is correct about this—and I can only say that it seems all too plausible—it is easy to imagine similar misconceptions about genetically engineered people. They might be viewed, irrationally, as preprogrammed, lacking ordinary kinds of human autonomy, and so forth.

Here, then, is one good reason for parents to hesitate to use either reproductive cloning or genetic engineering (though this is another case where the argument applies less well to PGD). This concern does not entail that all such parental actions should be morally condemned, however; it may depend very much on the particular circumstances. More to the point for this study, it is a dubious reason for invoking the coercive power of the state. The harms are not only somewhat speculative but also rather indirect. It is at least worth attempting to educate the guilty parties and forbid discrimination, rather than condemning what would otherwise be acceptable, and so making it even less likely that myths about genetically engineered people could ever be dispelled. Many parents rightly might be unwilling to have children who are at risk of social rejection, but some might be well placed to bring up their children within a highly accepting milieu. They might then lead the way for others, in which case the stigma will fade, as has happened with IVF (Pence 1998, 138).

Arguments for legal prohibitions *might* be stronger if genetic interventions involved some physically distinguishable mark, such as fluorescent skin, or webbed fingers and toes—given that such marks could easily attract social stigma. Yet even *this* basis for concern might eventually fade under some circumstances. Imagine, for example, a future scenario in which adults are capable of making similar changes to themselves and are not prevented from doing so. A time might eventually arrive when children with dramatic phenotypic differences would not be objects of misunderstanding or rejection. In any event, we are not confronted by a problem so urgent as to compel any draconian measures.

Conclusion

On reflection, concerns about the autonomy of genetically engineered children are largely misguided. That does not mean that they should all be dismissed entirely. For starters, some specific kinds of genetic modification could be detrimental to the development of a faculty of autonomy. Some attempts at modification of children's characteristics might be so comprehensive as to make the children feel like puppets or toys (and this ironically might also be the case with truly trivial modifications of appearance by genetic engineering). Some attempts at genetic modification might be part of an exercise aimed at excessive control of children's life plans. Even the more beneficial interventions *might* provoke adverse psychological reactions in the genetically engineered people themselves, even if these were based on a misunderstanding. Or they might trigger an unpleasant social backlash.

But when all that is said, the threats to the autonomy of genetically engineered children—and related threats to their welfare—have been greatly exaggerated and distorted by opponents of genetic engineering. Various smaller-scale harms are possible, but there is no reason to think that significant psychological and social impacts would be inevitable or even typical, or that they would have the enormous significance described by Habermas in particular. Some enhancements might actually increase the autonomy of the individuals concerned in the sense of boosting their powers of rational reflection, self-governance, and ongoing self-creation, expanding their available life options, and perhaps even increasing their capacity to scrutinize and revise their moral beliefs and understandings of the world.

5

Violating the "Natural Order"

An important fear that sometimes motivates resistance to emerging technologies is that of somehow sinning against nature, or violating, defying, or interfering with the so-called natural order. Kenan Malik (2000, 359–362) is one commentator, among many, who has drawn attention to the widespread hostility to biological manipulation, which he situates within a contemporary trend to value "the natural" over what is seen as artificial. This can be contextualized as part of an even broader cultural debate—over many centuries—between advocates of increased human understanding and control of nature, and others who look on such projects as impious, hubristic, or impermissibly defiant of the order of nature (e.g., Haynes 1994).[1]

In the specific context of human enhancement technology, J. Robert Loftis has identified four classes of arguments—what he calls "safety," "justice," "trust," and "naturalness"—that are commonly deployed against genetic engineering of the human germ line. He concludes that the first three classes, relating to harms, distributive justice, and our attitudes to the purveyors of technology, have some force, though no more than they have in respect to the genetic modification of nonhuman plant and animal species. On the account offered by Loftis (2005, 58, 72), it is the fourth class of arguments—contentions that he sees as lacking rational merit—that typically motivate opposition to genetic engineering.

To an extent, the motivation may be something rather different: status quo bias or fear of the unfamiliar. Yet Ronald M. Green connects this with a fear of violating nature or the will of God. Status quo bias can take the form of a belief that the current form of the human genome "represents the highest expression of human biological possibility," and this is sometimes given explicit theological support from an idea that the genome of Homo sapiens is divinely created. In that case, so it's thought, tampering with it is sinful (Green 2007, 8–9, 12–15, 171–176).

The current policy debates include their fair share of appeals to a supposed imperative to follow nature. It is often claimed that we must not "defy" the natural order, "interfere" with it, or "sin" against it. Such language is not used only by naive moralists and pontificating journalists. For example, Leon Kass's much-discussed opposition to human reproductive cloning is thoroughly permeated by language that invokes the naturalness and supposed profundity of heterosexual difference. As Kass (2002, 152) puts it, "Seeking to escape entirely from nature (in order to satisfy a natural desire or natural right to reproduce!) is self-contradictory in theory and self-alienating in practice." Throughout his work, there is an underlying assumption that we are beholden to a morally inviolable natural order of things. Similarly, Margaret Somerville (2007, 96) argues for an ethic that will privilege "the natural," although she acknowledges that in practice, it is extremely difficult to define its boundaries.

In this chapter, I will argue that appeals to the moral inviolability of nature are indeed highly problematic. I will then consider recent attempts by Richard Norman and Stephen Holland to rationalize the idea of violating nature. According to Norman's theory of background conditions, actions that are characterized by their opponents as violations of the natural order are better understood as threats to certain background conditions for choice that are widely accepted within the culture concerned. While this appears plausible as a descriptive and explanatory account—or so I will maintain—its policy implications are not so clear.

Holland claims that the fear of "violating nature" should be seen as rational and also relevant to public policy. I will argue the contrary: such fears should not be pandered to when liberal democracies formulate regulatory policy.

What Is Nature?

But what, in any event, is the natural order? What is our concept of nature, this thing that we must not defy, interfere with, or sin against? There are various ways in which we can think of nature, but there is no straightforward way to make a morally relevant distinction between the natural and unnatural. Writing in the eighteenth century, David Hume contrasted nature (in various senses) to the miraculous, the rare and unusual, and the artificial. He pointed out the difficulties of basing concepts of virtue and morality on any of these senses of nature (Hume 1985, 525–527).

Perhaps most famously, John Stuart Mill produced a sophisticated analysis of the issue in which he elaborated two conceptions of nature, essentially as follows:

1. Nature as the totality of all phenomena and their causal relationships, as investigated by science (Mill 1998, 4–5, 8).

2. Nature as those things that are not artificial—that is, whatever is not produced by human agency or technology (ibid., 7–8).

All such conceptions of nature, or the natural, lead us into philosophical difficulties. Most importantly, if nature is understood in Mill's first sense (which is similar to Hume's), nothing we ever do is unnatural. We are part of nature, and the same applies to all our actions, all sexual and reproductive practices, and every new technology that we can ever invent. Employing this broad sense of the natural, there is no useful distinction between natural and unnatural actions because *every* action is natural (ibid., 15–16).

On the other hand, if nature is understood in the second sense identified by Mill (equivalent to Hume's third sense), almost every action that we ever carry out is *un*natural. If unnaturalness is then used as a criterion for moral wrongdoing, it follows that almost everything we ever do is morally wrong. As for the other sense identified by Hume, which equates nature with what is statistically usual, it seems plain, as pointed out by Hume himself, that this cannot determine what is virtue or vice. Though moral sentiments are widespread, Hume (1985, 526–527) rightly observes, great virtue is actually as rare as great vice.

In short, none of the conceptions of nature discussed by Hume or Mill can help us distinguish between practices and technologies that are morally acceptable, and those that are not. Could this problem be overcome by an ingenious refinement of such conceptions? While I cannot locate the outer limits of human ingenuity, I see no reasonable prospect of modifying Mill's definitions in a useful way. One approach might be as follows. We could seek to refine Mill's second sense of nature by replacing the word *or*, in the definition above, with *and*. At the same time, we could define the term technology rather narrowly, perhaps restricting it to tools and practices that rely on advanced science. These adjustments to the definition, however, would have the effect that some acts commonly condemned as unnatural, such as homosexual practices, would be reclassified as natural and therefore morally acceptable.

That might not be a problem for anyone who considers homosexuality to be morally acceptable, but wishes to condemn various technologically based practices that supposedly sin against nature. For example, this modified criterion could be employed to condemn the contraceptive pill and IVF. Still, it obscures just what there is in common among the various practices and technologies that hitherto have been condemned for their unnaturalness. It also covers many things that are *not* usually criticized— at least not on the basis explored in this chapter—including computer technology, aviation, advanced building techniques and materials, and a vast range of medical therapies. Such a revision of Mill's second conception of nature would clearly not be a useful criterion for moral acceptability. Nor do I see how any other simple and useful change could be made to one or other conception of nature examined by Hume or Mill.

Perhaps, then, it is not surprising when contemporary moral philosophers give short shrift to arguments about what is or is not natural. For instance, Peter Singer notes that there is a sense in which it is natural for women to have children continually from puberty to menopause, but this does not mean that it is wrong to interfere with the process. For a utilitarian such as Singer (1992, 71–72), knowledge of natural laws, in the sense of regularities identified by science, can help us assess the consequences of our actions, "but we do not have to assume that the natural way of doing something is incapable of improvement." Similarly, Louis P. Pojman (2000, 123–124) offers a list of artificial innovations—from medicine and eyeglasses to "clothes, money, bicycles, cars, airplanes, houses, churches, and books"—that he sees as generally good. Even if it is thought that these authors move too quickly, attacking imaginary opponents, the issues raised by Hume and Mill may explain their impatience. They show the considerable difficulties that stand in the way of any theory that gives normative weight to a distinction between the natural and unnatural.

"Proper" Human Functioning

One alternative approach is to attempt to define the proper functions of relevant biological processes or organs of the body. Natural law theories that take a functional approach to human activities, bodily organs, and biological processes might be thought to avoid the problems inherent in the conceptions of nature described by Hume and Mill. For example, no appeal needs to be made to those conceptions of nature if we can find an independent and plausible way to characterize certain activities as failing to involve—or somehow impairing or suppressing—the proper functions of relevant biological processes or organs of the body. Such theories have a long history in Roman Catholic thought, and their doctrinal fruits are exemplified in the papal encyclical *Humanae Vitae* (1968), which addresses the moral acceptability of contraception.

Humanae Vitae condemns all acts of sexual intercourse that are accompanied by human actions (such as sterilization or the use of contraceptives) deliberately intended to prevent the possibility of pregnancy. This condemnation is based on the claim that there is a moral inseparability, established by God, of the procreative and "unitive" functions of sexual intercourse. Any deliberate attempt to suppress the possibility of procreation in a sexual act is thus a sin against God's design.

It nevertheless is not clear why we are morally obligated to act in accordance with a deity's design, unless it can be demonstrated that acting as specified is *independently* a requirement of morality (i.e., independently of the deity's purposes or commands). In any event, this must be demonstrated if such a theory is to have any prospect of convincing unbelievers or being taken into account in the formulation of public policy within modern liberal democracies. It is certainly not obvious that there is any moral requirement, independent of esoteric claims about what God (or a god) has designed and commanded, that (for instance) we must never engage in sexual intercourse while deliberately suppressing our fertility. It will not do to argue that this behavior would be unnatural in one of the senses identified by Hume and Mill, since such an assertion would be vulnerable to the criticisms already mentioned.

It might be suggested that a proper function of sexual intercourse is reproduction or procreation, so it is improper to suppress that function by deliberate human action. Yet this argument fails unless the word proper already includes a moral judgment. To rely on a claim that reproduction is a *proper* function of sexual intercourse, in the sense that it is a function that it is morally impermissible for us to suppress, is simply to assume what must be proved.

Evolutionary biologists recognize a sense in which the biological process of sexual intercourse in human beings in fact is *functional*, but the moral and political relevance of this is doubtful. Sexual intercourse and the human sexual organs are, I take it, adaptations: during the process of humanity's evolution, they tended to contribute to the inclusive fitness of the organisms involved. In many cases, the process of sexual intercourse continues to contribute to individual inclusive fitness (and as a side effect, the continuation of the species). We are not morally required to act in ways that are fitness maximizing, though. This would lead to absurdities, such as the conclusion that we must engage in sexual intercourse as frequently as possible when we are at our most fertile.

Once theological considerations are removed from the picture, it is more plausible that we should act in whatever ways are calculated to achieve our own conscious preferences, or whatever preferences we retain after reflecting on them rationally, while paying attention to the interests and preferences of other sentient beings. If it is claimed that our flourishing as individuals—something we presumably prefer—will be advanced by always acting in ways that are fitness maximizing or biologically functional for our species, this is just false. Many families flourish by our ordinary secular standards despite the fact that the parents used contraception to influence the number and timing of their children. Indeed, many individuals lead flourishing lives despite being sexually active and "childless by choice."

For reasons such as these, there are serious difficulties for any theory that attempts to explain the natural in terms of proper human functioning. The most that can be said for such theories is that they have their adherents in current liberal societies, and those adherents are entitled to advocate their moral views, act accordingly, and attempt to persuade others in noncoercive ways. All that duly noted, natural law theories that depend on a concept of proper human functioning are intellectually unconvincing, at least outside their traditional theological context, and are owed no more deference than other esoteric moral systems that form part of the social and ethical pluralism characteristic of modern societies.

The Theory of Background Conditions

Richard Norman developed a quite different understanding of the natural in a 1996 article titled "Interfering with Nature." This approach has since been elaborated and applied by Stephen Holland, who is more sympathetic than Norman to arguments drawn from nature. Norman's original article does not seek to impugn any particular practice. It actually defends IVF as a means of assisted reproduction. Norman does, however, develop an understanding of the natural that can be put to use by thinkers who oppose certain current or proposed biomedical technologies.

He begins, adopting an approach similar to Mill's, by pointing out some of the difficulties with the idea of defying or interfering with nature. Norman (1996, 3–6) then makes three main points that are essential for his redefinition of what interference with nature amounts to. First, our choices of actions, projects, life plans, and so on, are always made against a background in which some things are believed *not* to be open to choice. These "background conditions" for choice vary somewhat among different human cultures; typically they include general facts about sex, procreation, nurturing, maturing and aging, intergenerational relationships, death, the necessity of work, and the existence of illness and pain in our world. These are eternal verities of human experience, or so it might appear. Different human cultures develop their particular understandings of the background conditions to human choice, which are then categorized as the realm of nature.

Second, this understanding of *the natural* causes various paradoxical threshold effects. The fact, for instance, that the world contains illness and pain is a background condition to our choices to attempt to avoid or ameliorate them, in any particular case. Still, Norman argues, we do not wish for illness and pain to disappear from the world entirely. Without these things, many important practices would be lost, and our lives would lose shape, depth, and significance. Third, the risk is not that some specific harm will be done. The threatened elimination of basic, familiar conditions from the background of our lives instead creates the specter of a loss of experienced meaning. There is a sense, therefore, in which the continued presence of illness and pain as factors in human life is needed for our lives to continue to seem meaningful.

Norman suggests that the background conditions for choice are broad and general: not just any innovation will threaten people's sense of experienced meaning. Moreover, different cultures will understand the background conditions in different ways, and some may not be eternal verities at all, even though they appear to be so from within the culture. For example, many cultures have sexist beliefs about women among their most basic background "knowledge." At the same time, Norman observes, it is a psychological fact about human beings that we need a rather rich conception of the background conditions to our lives. It is, furthermore, not an entirely arbitrary matter what beliefs about the world are seen within a particular culture as the background conditions. Such beliefs are shaped not just by culture but also by our evolved biology and the physical world that we all live in. Thus we can expect a great deal of intercultural agreement (ibid., 6–7).

In Norman's view, the discomfort that some people feel about IVF and such prospects as biological immortality comes from a sense that crucial background conditions to choice—relating to procreation and death—are threatened. In this context, a "threat" to the background conditions appears to mean that certain familiar conditions may no longer obtain. A sense that these are under threat can be expressed as a claim that nature is being interfered with. Nonetheless, Norman (ibid., 10) defends IVF on the basis that incremental changes to our own culture's background conditions can be absorbed successfully into our thinking.

Norman has provided what appears to be a plausible explanatory theory that captures why certain technological innovations, but not others, are widely experienced as threatening. It also explains what these innovations have in common with such practices as homosexuality—practices that are not products of high technology but are also frequently impugned as being unnatural in a morally impermissible sense. According to Norman's approach, anything that might threaten a culture's basic assumptions about how ordinary human life works—especially assumptions about sex and its relationship to conception and birth, the development and rearing of children, male and female roles, and processes of aging and death—is likely be disquieting to at least some individuals. For example, homosexual practices might seem to threaten a background condition that relates sex to procreation.

If true, the theory tends to explain why apparently similar kinds of widespread psychological disquiet are produced by the contemplation of otherwise-varied practices and technologies, such as homosexuality, contraception, IVF, and PGD, and (more futuristically) reproductive cloning, genetic engineering, and radical life extension. It also explains why the apparently irrational idea of sinning against nature persists in debates about morality and public policy. It is worth noting, too, that the theory predicts at least some resistance to any new technology that seems to threaten the background conditions, even if the change offers benefits to some people, as with IVF. This is the theory's strength, as the prediction is consistent with historical experience.

Admittedly, those moral thinkers who have invoked concepts of sinning against nature have not usually expressed themselves in ways that are similar to Norman's description of a threat to background conditions. That is unsurprising, since Norman is not attempting to explicate actual arguments that have been made in the past. Rather, his theory implies that many assertions might be rationalizations of fears that are not properly understood by those who have them.

At the same time, some thinkers actually have developed arguments against certain technologies based on the claim that their availability would threaten the experience of meaning in our lives. The clearest case of this may be Bill McKibben's book *Enough*, which argues against the development and use of some radical kinds of technology that could change what McKibben sees as fundamental aspects of ourselves (or our children). These include our vulnerability to aging and death along with our genetic potentials for only limited levels of physical and cognitive ability. Attempts to "improve" ourselves by altering these fundamentals, as he sees them, would remove limitations that provide a necessary context for the experience of meaningful human choice (McKibben 2003, 46–47). This concern is close to the kind of fear that Norman believes we often feel, even if we are unable to articulate it.

In short, the theory of background conditions is an intriguing and suggestive contribution to our understanding of human nature. It should not be taken as conclusively established, but it provides a satisfying explanation for certain ideas about the inviolability of nature, and why they persist in modern societies, even though they are difficult to formulate and suffer great intellectual difficulties.

Holland's Contribution

More recently, Holland has defended appeals to the inviolability of nature when cultural understandings of the background conditions to choice come under threat. Holland (2003, 155) emphasizes the point that background conditions to choice are culturally specific constructs based on natural facts. An appeal to nature is, for him, a way of expressing hostility when the background conditions are threatened. Somewhat surprisingly, he believes that such expressions of hostility are *relevant* and *rational*, since they provide a way of defending conditions that are necessary for human beings in the culture to achieve an experience of meaning in their lives (ibid., 167–171).

Holland concedes that not all threats to background conditions will be perceived within the society concerned as morally unacceptable—at least for long. Where we recognize a need for some new technology, we are likely to accept it, even if we have initial doubts. We are unlikely to oppose a particular technology where to do so seems cruel or heartless, where the practices it enables seem relatively trivial (as with tooth fillings, though these also have great utility), or once the technology becomes familiar to us. In such cases, the technology's availability is absorbed into our thinking, and our understanding of the background conditions is adjusted (ibid., 169–170).

I have already commented on the explanatory power of the theory of background conditions as initially proposed by Norman. Holland's further elaboration has additional power. Consider the widespread opposition to homosexual acts that has existed in many human societies. This can be interpreted as, at least in part, the manifestation of psychological disquiet when a practice is perceived as threatening "the natural connection between sex and procreation" (ibid., 157). The persistence of a seemingly irrational hostility toward homosexuality might be explicable when the following points are considered. First, many conservative heterosexuals may find it difficult to see much utilitarian value in homosexual acts, since they may be unable to imagine how such acts can be genuinely pleasurable or how there can be anything unsatisfactory about the pleasures of heterosexual love. Second, for similar reasons, these conservative heterosexuals may find it difficult to understand the cruelty of moral norms or criminal laws that forbid homosexual acts. Third, questions of who has sex with whom do not seem trivial; rather, they have been troubling and fascinating in all human cultures for many thousands of years. Finally, although homosexual acts among human beings are doubtless as old as our species, they are still unfamiliar and seem confronting to many people.

Holland does not advance the kind of natural law argument found in *Humanae Vitae*, but no one—so he suggests—would want the emergence of a far-future society in which "all fertilization takes place without sex" (ibid., 167). He likewise explains the widespread unease about the genetic engineering of children by saying that it threatens our understanding of "natural and unnatural ways of promoting traits" (ibid., 157). More specifically, it threatens background conditions relating to choices and achievement in parenting: there are certain "natural" limits to what can be done to ensure the talents and other endowments of our children. We accordingly feel threatened by the prospect that too much of a technological guarantee would make parental nurturing seem meaningless (ibid., 186–188). Similarly, Holland alleges, human reproductive cloning is psychologically threatening for a combination of reasons. It couples the separation of sex from procreation with technological control of children's endowments (ibid., 157).

Holland looks at the theory of background conditions in the context of a detailed discussion of the ethics of genetic engineering, in which he demonstrates that a number of moral contentions can be offered in opposition to enhancement for, say, cognitive capacities. Some of the arguments are consequentialist in character, while others indicate that these efforts at enhancement would be morally wrong in principle. He concludes that the most powerful arguments tend to be consequentialist, while the assertions of principle are relatively weak (ibid., 184–185).

At this point Holland identifies what strikes him as a puzzle. Even if all the consequences were good, and even though the arguments in principle are so weak, "we would," he says, "continue to find the very idea of genetically altering our children for desirable traits such as intelligence and musicality, at least morally distasteful, and even morally impermissible." He suggests that this is because these interventions threaten a background condition for choice: that the endowments of our children should be, to some degree, out of our control and experienced as "given" rather than chosen (ibid., 185–189).

In fact, I am not at all sure that I would find the idea of genetic interventions to boost intelligence or musical talent troubling in the absence of adverse social consequences, and I take issue with Holland's repeated claims as to how "we" feel about these matters (or what "no one" wishes to see). Such language is, of course, almost unavoidable; often it is required, at least as shorthand, as the only alternative to producing clumsy prose. But sometimes, alas, it falls flat when a philosopher makes confident assertions about how "we" react emotionally or intuitively to some situation. In any event, "our" immediate, intuitive reactions may change when we are directed to changed circumstances.

Consider a case that Buchanan explores—that of seventy-year-old partners having children. While many people may have an intuitive reaction that this is repulsive, the effects will depend on the circumstances of the society concerned, notably "how vigorous 70-year-olds are like, how long they are likely to live, etc.," which could change considerably in the future (Buchanan 2011b, 128–129). If we contemplate older couples having children in altered circumstances, any intuitive reaction of repugnance that we start out with may not last.

All that aside, Holland's analysis suggests that in-principle opposition to genetic enhancement may not really be motivated by the reasons that are usually proffered in public debate. If this is so, how should we respond to it in the process of policy formulation?

The Argument from Background Conditions

Formulating the Concern
Should we accept Holland's claim that objections to the supposedly unnatural character of certain practices and technologies are actually relevant and rational? Imagine an opponent of homosexual acts, contraception, IVF, and reproductive cloning who has read the work of Norman and Holland. We'll call her Leonora. Leonora objects to all these practices, and advocates their suppression by the state (via criminal law or comparable methods) on the ground that they are sins against nature. When challenged, however, she retreats from this bald claim. Instead, she explains that what she really wants is to protect something that she

greatly values: the tight connection between sex and procreation, which she describes as a background condition to her life without which her own reproductive choices would no longer seem meaningful.

"Since I wish to protect a background condition without which my life will lose meaning," Leonora concludes, "my reaction is perfectly rational. It is also relevant in the sense that it should be given weight in the formulation of regulatory policy."

In fact, it is difficult to imagine that anyone would ever argue in precisely this manner; perhaps McKibben comes closest, but he is most concerned with dramatic changes to human capacities. The greatest horror for McKibben seems to be the prospect of technologically mediated immortality, which he imagines as time stretching ahead, "endlessly flat." He reports that this fills him with "the blackest foreboding"; he sees it as likely to take the joy and meaning from life, making it "melt away like an ice cream on an August afternoon" (McKibben 2003, 160–161).[2] Later in this chapter, I'll return to some of McKibben's specific points.

The argument that I've attributed to Leonora does not depend on such dramatic possibilities as genuine immortality. All that it requires is some innovation or some existing practice such as homosexual acts that seems to threaten what she perceives as background conditions. Her contention has an immediate appearance of wrongheadedness, although it is difficult to define exactly why. It stems, perhaps, from Leonora's third-person account of her own feelings of meaningfulness; as I've characterized her, she seems too sophisticated about such things to be sincere. At the same time, part of the difficulty in refuting her consists in the fact that she makes a point with a degree of force: given the way she is psychologically constituted, she actually *may* suffer distress from the knowledge that homosexual practices, IVF, sex with contraception, and reproductive cloning are going on in her society, even behind closed doors.

Given that she has such emotional reactions to these practices, perhaps it is prudent of her to oppose them and seek their prohibition. In that limited sense, at least, her hostility is rational: she is trying to prevent an outcome that she genuinely fears. Holland himself reacts somewhat like Leonora when he discusses the genetic enhancement of children's traits. As noted above, he suggests that "we" would retain our moral distaste or opposition to the genetic enhancement of children even if all the consequences were good.

Leonora would doubtless obtain further support from Michael J. Sandel (2007, 45), who objects to "designer babies" on the basis of an "ethic of giftedness," which values the unpredictability of children's qualities and (in words that Sandel attributes to William F. May) an "openness to the unbidden." Such enhancement efforts would, Sandel later adds, entrench a stance of mastery and domination toward the world—a stance that fails to appreciate what he calls "the gifted character of human powers and achievements," and misses "the part of freedom that," he alleges, "consists in negotiating with the given" (ibid., 83). Thus, Sandel relies on specific concepts of morality, virtue, and freedom, but it is unclear why the state should exert its coercive power to take his side on these contentious points. If the state were to take sides in such matters, why not side instead with the idea that attitudes such as Sandel's are imprudent and defeatist (cf. Pence 1998, 136)?

Perhaps, indeed, the idea that children's characteristics could be *completely* controlled is frightening. This is partly because of the degree of responsibility that this would place on the shoulders of parents, and perhaps because there can be value in an element of surprise (depending, to be sure, on the *kind* of surprise). There is also the residue of concern identified in chapter 4: as the parents' degree of genetic management increases, children might come to feel like mere puppets. None of this, however, establishes that political coercion should be used in an attempt to prevent any attempts at all to influence children's genetic potential for certain traits.

A Liberal Response

Should the rest of us support Leonora? No, we should not. It certainly is easy to understand how an individual such as Leonora, who is strongly impressed by the sex-procreation link, might find something troublingly anomalous about nonprocreative sexual acts or the use of reproductive technologies that can enable children to be born without acts of vaginal intercourse. Yet the existence in her society of any number of such anomalous acts will not, by itself, prevent her from enjoying as much "normal" sex as she wishes. That is something of value to her that she *won't* lose. Nor does it prevent her having children, enjoying all the activities associated with raising a family, and generally living a meaningful life (by her own standards). Nothing plausible that is said by Holland entails that Leonora will *actually* cease to experience meaning in her life if homosexuality, IVF, and so forth, are allowed by law—human beings are surely more flexible and resilient than that.

Consider the fact that many people are already engaged in homosexual acts every day. This does not prevent someone like Leonora from leading a more conventional way of life, or finding it deeply absorbing and valuable—subjectively worthwhile.

If the jurisdiction where Leonora lives provides a system of same-sex marriage, it might be said that the *exact cultural significance* of heterosexual marriage will change: it will no longer be the only form of sexual relationship that is given a particular kind of official recognition. I am tempted, though, to ask whether this is really a mere "Cambridge change": a supposed change that leaves the intrinsic properties of the thing as they were. After all, in the example given, the actual practices will not alter. Rather, Leonora wants the practices to be *regarded* in a particular way by her society as a whole. For the sake of argument, let us concede that she has an interest in this, and likewise if some people begin to use reproductive cloning to conceive children. This will not undermine the plans of couples who continue to have children in the usual way, but it might affect the exact range of cultural significances that their society attaches to sex.

Even if we regard these as real changes, they do not mean that heterosexual married couples will experience a loss of meaning in their lives. Nor, in a scenario in which some people make use of reproductive cloning or other enhancement technologies, need other people find that their children's nurturance as well as the rest of their lives' activities no longer seem meaningful or worthwhile. These more conservative folk will not thereafter find their relationships pointless, futile, or lacking in personal significance and value; their lives will not be experienced as absurd or worthless. As time goes on, they can continue living lives that they find rich and meaningful, even if they fondly wish that their society gave its exclusive support to their particular way of life and conception of the good.

It seems, then, that the most that Leonora can plausibly argue is that certain changes will make life *less* meaningful for her, although we might be skeptical about what that really amounts to. For example, will her erotic and familial activities now be only 90 percent as satisfying? More plausibly, she will simply lose the benefit of living in the kind of society that she wishes and in which she feels most comfortable. I acknowledge that this might not be an entirely trivial loss. Expressed this way, however, the contention becomes no more than a variation on a familiar class of illiberal arguments discussed in detail by Feinberg (1988, 55–63)—a kind of impure moral conservatism based on harms to interests.[3] It then suffers the familiar problems of all such claims in that it advocates an attempt to freeze social change and values the interest in controlling what kind of society one lives in above others' interests in exercising individual autonomy.

There is, of course, always the possibility that a future society will be dramatically different from our own. The long-term effects of change may produce a social environment that we would find strange indeed, if we were forced to live in it. The society that eventually comes to be may not have marriage-like relationships at all, or it may achieve the total separation of procreation and sex that troubles Holland, and would surely trouble Leonora. Gregory Stock, an advocate of human enhancement, offers a contrasting reaction. Stock acknowledges that the future may be one that we (or some of us) would find alienating, as our great-great-grandparents would probably find the customs of today. Yet he adds that it will not be experienced like that by those who actually live in future societies: "A few romantics may one day look back on our era as a golden one, but far more future humans will see today as a primitive, difficult time, far inferior to the world they know" (Stock 2002, 150).

While this may seem like an overly confident speculation, it is readily imaginable that future people will live happily in a social environment that many currently living people would find distressing—assuming that quite different problems such as global warming and the depletion of natural resources do not intervene to prevent this happy outcome. While much may change, there is no reason to believe that the people in the future society will lack involvements in various kinds of relationships that they will find worthwhile. Even if Leonora does not approve of such a society, and dislikes thinking about it, that is likely to be because of her particular moral beliefs rather than because she can point to any kind of secular harm that people in the future society will have to endure.

The future people described by Stock may lead a strange way of life by our standards, early in the twenty-first century, but there is no need for us to imagine them as *suffering* or think of the outcome as inherently evil. Nor does the possibility that these future people will one day come into existence cause Leonora any harm, above and beyond her feelings of disquiet and moral indignation when she thinks about it.

Admittedly, there may be some limit to how quickly individuals can adapt to truly rapid and sweeping social and technological changes. If enough changes happen at a sufficiently great pace, some people may merit our compassion as they come to feel like aliens in their own land and strangers to their own families (Feinberg 1988, 48–49). But of course this is already a familiar problem to many people as they age, while technological, social, and economic conditions change around them. Only draconian, highly illiberal actions by the state could prevent or substantially hinder this situation. Any decision to do so—or to try—must be made with great reluctance. I will return to this point but note, too, how rapid social change is frequently dampened by conservative reactions, even without any draconian exercise of political coercion. For example, as Pence (1998, 67–68) observes, the sexual and cultural revolution of the 1960s led to something of a conservative reaction (even before the AIDS outbreak began).

In any event, not all background conditions are similar in their demands. If, for example, it is a background condition to many social institutions and activities that there is *some* illness and pain in the world, then the immediate abolition of all illness and pain might be something that we would be psychologically unprepared for. Nevertheless, no foreseeable medical advance is likely to threaten a background condition of *that* kind. If such an extraordinary change ever happened, it would surely be in a future society far different from our own, with people whose psychological preparedness might be quite different from ours. For better or worse, we are not confronted by a choice between the benefits of a supermedicine and whatever psychological comforts we would lose if we developed it.

Again, while we may not be psychologically prepared for a world where the endowments of our children are totally under our control, there is no prospect that we will soon experience it. For the foreseeable future, any capacity that we might have to enhance the abilities and prospects of our children by manipulating their DNA will undoubtedly have severe limits, much like our capacity to manipulate our children's environments for the same purpose.

Unless they are truly rapid and sweeping—as with an invasion by a technologically superior civilization—social changes that alter the exact cultural meanings of practices can be absorbed, psychologically, as they take place. Fears of the kind that we have attributed to Leonora are steadily overcome, innovation by innovation, and this piecemeal process is all to the good. It allows many kinds of people with a wide range of values and interests to flourish together in the same society.

As Feinberg argues, gradual changes, even to ways of life, are virtually inevitable, but it is not reasonable to analogize them to acts of invasion or genocide, the violent suppression and obliteration of a religious faith, or the abrupt extinction of an animal species by catastrophe or overhunting. Where change is incremental and unforced, it is fairer to compare it to the development of languages over time or the evolution of biological species. Even if something is lost, something is also gained, and those at the end of the process will have no grievance against those who produced the changes along the way (Feinberg 1988, 68–82).

In all the circumstances, then, it is difficult to see why we should be solicitous about Leonora's *welfare* (as opposed to her pretensions of virtue or moral rectitude) to the point of deferring to it in public policy.

As new innovations are introduced, and at least some of them are taken up by others, her society may well change around her in ways that she finds distressing. If public policy gives effect to her views, however, many others will be prevented from living as they please, shaping their lives by making their own choices about personal and significant matters. An attempt to freeze the specific cultural significances of certain acts in an effort to please some people would severely restrict how other people live their lives. Leonora, it must be remembered, is insisting that *all* sexual acts must be potentially procreative and that *all* human reproduction should be the result of acts of vaginal intercourse. If she reflects, she should understand that she is being, in a familiar sense, unreasonable—she lacks the important moral virtue of being prepared to accommodate the values of others.

More generally, the experience of life in modern societies that embrace social and ethical pluralism casts doubt on the idea that an entire society must adopt the same rich set of background assumptions. Whatever may have been the case in tribal or pretribal societies, or even the smaller societies of ancient, medieval, and earlier modern times, twenty-first century societies accommodate great variation in people's worldviews and values, including what different individuals and groups experience as natural. The citizens of modern liberal democracies are perhaps psychologically unprepared to act with no limits at all, as if they were omnipotent and omniscient deities; the idea of choice in that circumstance can produce a feeling of vertigo if we try to imagine it. But no conceivable technological advance will bring about such a situation.

It is important to be clear about the nature of this argument. Leonora might respond to everything said up to this point by insisting that she remains firm in her moral judgments, and that these are not falsified by the fact that others do not share them, the fact that they might not be widely shared in some other societies, or the expectation that they will not prevail. It will not impress her if she is told that the tide of history is against her, and that people living in, say, the twenty-second or twenty-third century will not share her concerns. It might not even matter to her that she can imagine *her own* beliefs about matters of morality changing (she could accept that this is possible or even probable, but construe it as the result of malign influence from a decadent social environment). None of these considerations counts against her moral belief or judgments, or gives her a good reason to discard them.

But that is not the point. It is true that Leonora is not rationally compelled to abandon her moral position, and she may continue to put forward arguments in its favor in an effort to persuade others. Still, she needs to claim more than this if she wishes to argue for legal prohibitions of practices that she dislikes, since her society allows people to live by a range of personal moral positions. She needs to disentangle her moral objections to cloning, homosexual acts, and other practices that she dislikes from her claim to be *harmed* by them in a sense that is relevant to public policy. Unless she finds a more compelling case, we cannot accept that she is wrongfully harmed, merely because her efforts to influence and persuade others prove to be unsuccessful, and her society changes in ways that are contrary to her preferences.

The Role of the State

What should be the outcome if Leonora is part of an electoral majority that supports laws prohibiting all the practices that she dislikes? This relates to the more general question of what should happen if an electoral majority or government in power wishes to forbid acts that, it might be said, do no secular harm, but may undermine or tempt people away from certain ideals; Gerald Dworkin (1988, 163) offers the example of chastity. But what if Selene, as I'll call her, is dedicated to carnal pleasures? Why, except that she is in the minority, might she not seek to have the state silence those who would disparage or discourage her activities (perhaps by tempting away potential lovers to a life of ascetic religious practice)? If Selene's political party manages to obtain the electoral numbers, should it attempt to suppress whichever organizations advocate "chaste" attitudes to sex? Is the choice of which sexual ideals will be protected by state power to be made an electoral issue?

If liberal democracies are genuinely liberal, their overriding political ideal is a public policy based on mutual tolerance. It is not sufficient for someone like Leonora to claim that somebody else's actions, while causing her no overt harm, are inconsistent with the way she would prefer the world's background conditions to be. If we are all to flourish together in the same society, she and others ought to be adaptable to the ever-changing conditions of the world. If we are going to coexist, then all of us will have to give up our attempts to use political coercion to control exactly what cultural meanings are assigned to our respective ways of life.

Despite the undoubted intellectual value of his contribution, then, Holland is too indulgent toward what are plainly illiberal reactions. Indeed, the theory of background conditions gives us reason to be more suspicious of arguments against the acceptance of new technologies since it suggests that an apparently rational gloss may be laid on essentially irrational or at least nonrational fears.

McKibben's Arguments

My response to the imagined Leonora also provides a general answer to McKibben's claims about the loss of meaning. When a new technology is first developed, it may appear that the sense of meaning in our lives is under threat; however, we do adapt, and new social practices incorporate the new technology. As long as technological change does not proceed too quickly for adaptation, arguments based on the loss of experienced meaning should carry little weight for public policy.

Nonetheless, it is worth pausing to consider at least some of McKibben's specific arguments to ensure that they are not dismissed too quickly. As already mentioned, McKibben is concerned with dramatic changes to human capacities. Apart from the prospect of bodily immortality, brought about by technological means, he imagines situations in which we are invited to see the individuals concerned as genetically engineered so thoroughly and successfully that life no longer poses them any meaningful challenges. Or alternatively, we are invited to see them as having lost a sense of their own identity. Typical examples in his book *Enough* are the runner who feels no weariness from the strain and the genetically engineered rock climber who wonders, "Is it really me doing this?" (McKibben 2003, 53).

Elsewhere, McKibben examines the prospect of technologies that will give us, he envisions, seamless articulation with powerful machines and unprecedented control over our environments. His response to such a prospect is despairing. He suggests that such technologies would enhance our power for no meaningful purpose and culminate in a kind of suicide of our species. They would, so he alleges, be deadening and muffling, cutting us off from the world with which we coevolved (ibid., 93–99).

All of this, however, is unconvincing. It is inconceivable that any technology will remove all challenges from life; that would require omnipotence. If our powers of action were massively increased, either by direct changes to our bodies or the availability of increasingly powerful tools, we would doubtless be able to accomplish far more than is currently possible. Yet there would still be areas within which we would be challenged, neither sure of succeeding nor doomed to failure. Instances of a life without challenges are impressive only if we conflate a lack of *any* challenges with the easy ability to overcome some of our *current* ones. Responding to Sandel's concerns about a quest for mastery, Buchanan (2011b, 80) observes that there would still be plenty of limitations even in a society where people live for four hundred years:

People would still die of accidents; wars presumably would still occur; deadly pandemics presumably would still arise; people would still fall in love with people who do not love them and fail in every effort to make themselves loveable; children would still sometimes wound their parents by repudiating their values or failing to appreciate their sacrifices; people would still invest in careers and projects that fail, despite the best laid plans; the weather and natural disasters, would still be beyond our control; many human actions, both individual and collective, would still have unpredicted consequences, etc., etc.

It should also be said that the rock climber's question is not coherent. Since *none* of us are self-creators all the way down, as discussed in chapter 4, none of us can distinguish a "real me" from the empirical me that is ultimately a product of external causes. The question makes no more sense when asked by genetically engineered rock climbers than when asked by ordinary ones. A genetically engineered rock climber might be able to scale more difficult cliffs than others, but only if they have taken advantage of their genetic potential by adding hard training and the mastery of technique. There should be no doubt that the real "them" is meeting the challenges, exactly as much as if their genetic potential had been purely the result of chance rather than partly the result of choice.

McKibben is especially concerned with the possibility that advanced genetic and other technologies will lead to immortal lives, though it is difficult to pinpoint exactly what he finds so appalling about this. Why, in particular, is he convinced that immortals would find their lives lacking in joy and meaning? It is no doubt true that immortals would experience life *differently* from current people, who face the inevitable prospect of old age, frailty, and death. But that does not entail that immortals would experience their lives as pointless, futile, joyless, or worthless, or that they would live without finding personal significance and value in their activities as well as relationships.

In one passage, McKibben (2003, 160–161) offers a number of reasons for pessimism. He writes of a "divorce" between immortals and the rest of creation, although this seems to mean no more than that immortals would cease to be part of a system in which every living thing dies and gives up its nutrients to the land as well as new living things. While McKibben may place a quasi-mystical value on this process, it does not follow that the immortals themselves would feel joyless or find life meaningless. McKibben writes, furthermore, of a divorce of each human individual from every other, but again it is not clear what this really amounts to. The complaint seems to be that immortals would not owe care to their juniors and attention to their elders, yet he does not explain exactly why.

There would be at least some children, and they would still need nurturing and guidance. It is true that a society of immortals would not have elderly people suffering from frailty merely as result of old age, but surely that fact alone would not make life for immortals joyless or meaningless. Why would it not be an *improvement*? Might there not even be a special joy, in such a society, in the interactions of parents and children while both are in the prime of their adult powers? The immortals might look back on many aspects of our own situation with pity. In any event, I see nothing in McKibben's analysis that goes anywhere near compelling a belief that the immortals themselves would experience life as stale.

Imagining the Future

As I've argued elsewhere, it is difficult to imagine what things might be valued in a future society far different from our own. We can't predict what would give its inhabitants joy, and a sense of connection with each other or other elements of their world, or what they might experience as meaningful. We can elaborate vividly on the things that we stand to lose as the future emerges. We are well equipped to identify practices and institutions that might be transformed or lost. But we cannot imagine the detail of what might replace them, and can think only in general terms of what we stand to gain (Blackford 2011–2012).

Agar (2004, 62) makes a similar though narrower point in his book *Liberal Eugenics*, where he asks us to contemplate "the moral image of food finding." He emphasizes the difficulty, danger, and investment of time that was once involved in obtaining and preparing food:

The hunting of certain animals required teamwork and entailed a risk of death; crops would be planted only to fail. Technology has transformed the significance of food and of the methods historically used to procure it. Although many people fish and hunt recreationally, these activities are travesties of the fishing and hunting of our ancestors. We get food in other ways. (ibid.)

When we employ those other ways, they have little of the significance of hunting for our prehistoric forebears. For us, it is more or less meaningless when we put a jar of pasta sauce in a supermarket cart (ibid.).

But is this an example of meaning being taken away? Hardly. Agar correctly adds that technology has displaced rather than destroyed the significance of food finding. We prepare beef Wellington for our guests rather than informing them, "I have meat." We devote saved time from hunting and planting to reading novels, watching movies, reflecting on life's meaning, or going fishing, which, as Agar (ibid., 62–63) observes, "is usually more about spending time with nature than bringing it home for the cooking pot."

From all this, Agar (ibid.) deduces that we should not rush to condemn a technology simply because it severs activities and satisfactions from what was customarily required to enable them. Yet there is a broader lesson here. Our distant ancestors were in no position to predict modern ideas of food acquisition and cuisine, and nor could they have predicted much else about our modern societies, including what activities and experiences now provide us with deep satisfactions, or a sense of joy and meaning. By extension, we are ill equipped to predict what activities and experiences will be satisfying, joyful, or meaningful for future people who might grow up and interact in environments quite different from our own, and who might have very different experiences, understandings, and skills. It does not follow, of course, that such people would lack the experiences of satisfaction, joy, and meaning in their lives.

That said, Agar (2010, 1) himself has more recently argued against what he calls "radical enhancement," which "involves improving significant human attributes and abilities to levels that greatly exceed what is currently possible for human beings." His general line of reasoning is not that our lives or those of our children would be miserable, or lacking in satisfactions, joy, and meaning. This is important, because it leaves us no real basis to feel sympathetic concern for radically enhanced people. The problem, rather, is that they would have lives that are lacking in elements that are valuable by our current standards, or by what Agar thinks of as *human* standards.[4]

Accordingly, Agar advocates a form of species relativism about values, and wishes us to judge future lives by what he takes to be the standards of Homo sapiens. Judged by (what Agar views as) these standards, radically enhanced people would lead impoverished lives, even if their lives were full of satisfactions, experienced as joyful and meaningful, and so on. Agar imagines the destruction of various institutions that are important to us, such as marriage and sports as we know them, in a world populated by near-immortal people with extraordinary physical and cognitive abilities. He seems to consider this morally unacceptable.

While this approach is ingenious and merits our respectful consideration, it applies only to truly extreme enhancement of capacities. For example, Agar envisions people who are not merely very longed lived but almost immortal in the sense that they never grow old. Accordingly, much of what is contemplated in this book remains unaffected. In any event, I am skeptical. For instance, consider how we might reply to a prehistoric hunter who regards our current lives as impoverished—judged by what the hunter imagines to be human standards.

One possibility is that we might eventually persuade the hunter that by the standards that *really* matter (the genuine human standards, if you will), our lives are not impoverished after all. Those standards might relate to the experience of joy, meaning, and so forth, or they might relate to quite different things such as our ability to act on the world or the complexity of our experience. Depending on which standards are chosen, our lives might seem no more impoverished than the hunter's—indeed, they might appear *less* so. If the hunter does not acknowledge this, we might attribute it to a certain narrowness of thought, perhaps the result of socialization. Why shouldn't radically enhanced people take a similar attitude to us if we don't accept their criteria?

In fact, the example of the prehistoric hunter suggests that if we could apply our own deepest criteria, whatever they really are, we might judge our own lives as inferior to those of the radically enhanced humans, or at least as no better. If we judged the radically enhanced humans' lives as impoverished compared to our own, this might show a lack of imagination holding us back.[5]

Conclusion

This chapter's analysis indicates that there are considerable difficulties in constructing any sound argument for a political rejection of a practice or technology on the basis that it violates the natural order. Certainly, any such claims should be examined with considerable skepticism, given the plausibility of the theory of background conditions, which suggests that some individuals will react to a sense that basic conditions of their world are being changed in a disorienting manner. These people may have an underlying (though probably unjustified) fear that their lives will no longer seem meaningful in a world where certain practices or technologies are allowed.

The theory of background conditions gives us an additional reason for skepticism when we see illiberal attacks on enhancement technologies. Our skepticism should be heightened when appeals to nature appear in the mix of arguments against a particular practice or technology. Of course, the strength of any genuinely compelling arguments cannot be removed merely by the fact that a *bad* argument is *also* run by opponents of the practice or technology. That granted, where appeals against violating nature form one element in public debate about some innovation, this should sound an alarm. It is likely that opponents of the practice or technology are, to an extent, searching for ways to rationalize a psychological aversion to conduct that seems anomalous within their contestable views of the world. In such circumstances, it is also possible that this aversion will influence some legislators and members of the public.

This chapter and the previous one have moved from concerns about direct harms to individuals to the possibility of more indirect and intangible harms. Sometimes the claims made by opponents of human enhancement involve harm to society as a whole, as with the (admittedly implausible) claim by Habermas that genetic engineering is incompatible with liberal society. It is time to focus more specifically on claims about indirect and intangible harms.

6

Indirect and Intangible Harms

This chapter discusses the circumstances in which enhancement technologies might undermine social harmony and stability, or have other effects that could be described as harmful to society as a whole.[1] The latter idea should immediately raise some suspicions. Liberal democracies cannot, for example, accept that harm has occurred merely because the citizens have adopted a way of life that fails (according to one or another theological viewpoint) to facilitate spiritual salvation. Nor can it count as harmful if a particular moral viewpoint, way of life, or conception of the good—among the many that the society tolerates—fails to maintain its popularity.

Then there are outcomes that we might be tempted to consider harmful, but only from a specific perspective. Perhaps, for instance, a social practice changes to such an extent, and in such a manner, that it can be described as decadent. This might seem harmful—but who is harmed by it if no individuals *experience* it as harm? Or the practice might remain available only to a select group, while losing its popularity among the masses. Again, who is actually harmed by this? However unfortunate such outcomes might appear from one or another vantage point, it is not clear that they are genuine harms at all, or how a claim that they are harms to society as a whole is more than a metaphor or contestable value judgment—perhaps made from some perfectionist viewpoint. At most, these outcomes seem like free-floating evils that harm no individual's interests.

Indeed, Feinberg distinguishes between actions that are harmful or offensive, on the one hand, and those that do not cause harm or offense but do produce impersonal or free-floating evils, or alleged evils. According to Feinberg (1988, 4), the most obvious of these (alleged) evils are drastic social change, exploitation that has been consented to by its victim, and degraded taste. There might be others, however, such as the loss of certain kinds of social diversity, certain moral perspectives or ways of life, or certain heights or depths of experience. Another possibility is that individuals will be caused to act in ways that violate a moral norm that one group or another considers an element of the "true" morality—although nobody is thereby harmed.

In this chapter I look at such alleged evils only briefly, since I generally accept Feinberg's view that they provide only weak reasons for the suppression of conduct by means of political coercion. My emphasis instead will be on harms that might arise as the indirect effects of enhancement technologies. In some cases, the harm might crystallize, as it were, only from many people's cumulative actions.

It does seem possible that the spread of some new practices could bring about social effects that might reasonably be thought of as harm to society as a whole. Such practices might, for instance, reduce the society's pool of available talent or undermine its stability. Accordingly, my main focus in this chapter is on innovations that might have such effects. Considerations of social stability in particular suggest that we could employ an expanded principle of intangible harm, recalling the ideas advanced by Lord Patrick Arthur Devlin during the debates in the United Kingdom about prostitution and homosexuality in the late 1950s and early 1960s. As a matter of abstract theory, such a principle might justify certain legal restrictions on human enhancement. I will argue, however, that any such principle brings its own dangers.

As in previous chapters, a residue of concern remains after the issues have been considered thoroughly. But this needs to be balanced against the reluctance of liberal societies to violate the autonomy of individuals.

Harms and Evils

Feinberg distinguishes between what he calls pure and impure forms of legal moralism. The pure form seeks to prevent immorality or evil outcomes for its own sake, while the impure form is focused on further consequences of these things—that is, harms such as those involved in the disintegration of a society. Feinberg makes an additional distinction between strict and broad forms of pure legal moralism: the former seeks to prevent immoral actions as such, while the latter seeks to prevent them for the purpose of preventing evil (but nonharmful) outcomes. Finally, he distinguishes strict and broad forms of impure legal moralism: the former classifies the actions that it seeks to prevent as inherently immoral, while the latter does not, but both ultimately rely on the indirectly harmful effects of those actions to justify prohibition, making them impure (Feinberg 1988, 8–10).

This set of distinctions is conceptually useful, though it can be difficult to assign individuals or viewpoints in a clear-cut way. For example, Feinberg categorizes James Fitzjames Stephen (1993, 12–15) as a pure legal moralist, whereas Devlin was an impure moralist, since his arguments were ultimately directed at preventing social breakdown. On balance, this is a reasonable analysis of two classic positions identifiable within the extensive literature. Consider Stephen's views, though: Stephen in fact did maintain that certain acts should be forbidden purely because of their moral wickedness; his work nonetheless also contains impure elements, since he argued for the use of force to establish and protect the institutions of religion and conventional morality, in part because of the good supposedly achieved by them (ibid.).

Conversely, Devlin's work is open to interpretation. On one approach, he might be seen as adopting an already moralized concept of the survival and harmonious functioning of a society, in which the spread of so-called immoral actions itself counts as a loss of social functioning, whether or not it leads to more obvious kinds of disintegration such as a collapse into disorder. If, however, this is the only sense in which society has broken down, it is difficult to see why we should resist it by political means. Liberal democracies tolerate a wide range of values and ways of life, and do not attempt to suppress actions merely because they have traditionally been considered immoral—and might, if not suppressed, become popular. In any event, Devlin is usually seen as presenting the *social disintegration thesis*, as Feinberg (1988, 43) calls it, or the idea that unsuppressed immoralities will lead to chaotic conditions that are harmful to all those involved.

The distinction between harms and evils is not entirely clear, and the characterization of an outcome as a harmless evil will often involve controversy. Consider, for one, Somerville's concern that enhancement technologies could lead to the cumulative outcome of a society in which no one suffers from Down syndrome, profound deafness, or bipolar disorder, and no one is achondroplastic or homosexual.[2] Even if this came about noncoercively as a result of many individual decisions, it should never be tolerated in public policy—so Somerville (2007, 131) claims.

Yet this provokes questions. First, why is such an outcome a matter of regret? Is it because somebody has been harmed? That seems implausible: it is difficult to identify individuals who have experienced harm in such a case. Presumably, the various individuals in the future society will not *mind* being heterosexual, or feel that they are somehow *missing out* by not suffering from Down syndrome, bipolar disorder, and so on. None of these people, as we attempt to imagine their situations, seem like good candidates for our compassion. If the absence of Down syndrome, for instance, is considered an evil, it looks like a free-floating one, in Feinberg's terminology, but it is not apparent that Somerville (ibid.) sees it that way, since she laments that it would mean the elimination of the "special gifts" that Down syndrome sufferers bring to society.

From that viewpoint, it seems that the aggregate outcome is harmful: the citizens of the future society will need to live without these gifts, whatever exactly they are. C.A.J. Coady (2009, 171–172) suggests that emotional warmth and a lack of inhibition are the valuable qualities that inherently accompany Down syndrome, but many other people who do not suffer from Down syndrome also possess these qualities. Perhaps another way to look at the issue is to lament a loss of diversity of certain kinds, although one might wonder whether diversity in itself, whatever form it takes, is an unequivocal or uncontroversial good.

Another possible evil that is difficult to distinguish from harm is the type of outcome in Aldous Huxley's (1932) *Brave New World*, where much that we currently value (art, science, and certain kinds of relationships) has been lost. Even though the citizens of the brave new world feel subjectively happy, they live what we are invited to see as impoverished, psychologically flattened lives. The complaint even has a Millian ring: the society depicted in Huxley's novel can be described as one in which "higher pleasures" (Mill 1910, 11) have been lost. Elsewhere, I have specifically argued against the claim that the availability of enhancement technologies would result in the kind of psychological flattening that is so obvious in *Brave New World* (Blackford 2003; 2005). What if I am wrong? If some kind of sociotechnological slippery slope led to *this* scenario, might it be reasonable to think that society as a whole is harmed, and in any event might it justify the passage of legislation designed to head off such an outcome?

Similarly, from a perfectionist viewpoint, Thomas Hurka has critiqued large economic inequalities. He fears that perfectionist values could be undermined in several ways: large inequalities can frustrate cooperation and the good that this can bring as well as inhibit personal relations; they can also direct ambitions toward the attainment of wealth rather than the achievement of perfectionist goals (Hurka 1993, 176–183). Might something similar apply to the pursuit of certain kinds of enhancement of genetic potential? A perfectionist element is present, too, in Sandel's concern that practices such as sports and entertainment can be corrupted by technology; for example, according to Sandel (2007, 29–44), US football is corrupted by the presence of gigantic linemen fed on a super-rich diet, while light opera has been corrupted by the use of sound amplification.

There may be other, even more obscure and speculative, social outcomes that could be viewed either as harms or free-floating evils. Some of them seem particularly dubious, however. Holland, for instance, discusses whether the widespread enhancement of musical ability would undermine the value placed on musical ability by the society concerned. Yet it is unclear how anyone can be *harmed* by this enhancement or why it is an intrinsically evil outcome if future societies, in different circumstances from ours, place value on different things. In any case, why wouldn't a wider spread of innate musical talent actually lead to more people having a discerning appreciation of music? Holland (2003, 183–184) concludes, I think rightly, that hostility toward genetic enhancement is unlikely to be based on a sense of possibilities like these.

Each of these scenarios could be examined in depth, but the main point that needs to be made is that they all provide dubious bases for public policy. Take the assertions made by Somerville. It is an implication of her stance that the state may, and where necessary must, use political coercion to ensure that there are enough children with Down syndrome, bipolar disorder, profound deafness, achondroplasia, and homosexual leanings to serve the needs of others by providing their gifts. No doubt some people who fall into the classification that she mentions do have socially valuable gifts, but so might the children who would be born if a parent chose, say, to implant an embryo *without* the genetic potential for Down syndrome or bipolar disorder.

As usual in Somerville's work, what makes it positively *illiberal*, not just morally conservative, is the role she assigns the state. Somerville, of course, has every opportunity to persuade us individually to reject such things as selective abortion or the use of PGD to avoid traits like deafness or achondroplasia. She goes further in seeking that such conduct be stopped by the use of political coercion.

If the argument is conducted in terms of lost diversity, we might wonder whether *all* sorts of diversity should be valued. If so, this leads quickly to some unpalatable conclusions: Do we really wish that we lived in a world in which some people still suffered from smallpox? Will it be a matter of regret if we ever wipe out malaria, thus leading to an absence of malaria sufferers and whatever special gifts they might bring to the world? Do we want to ensure that a certain number of people suffer from cystic fibrosis? What about Huntington's disease or bowel cancer? If it is somehow a bad thing to eliminate profound deafness, what about congenital blindness, or the combination of both? It may be rational to value some kinds of diversity, if they bring joy and color to our lives, but valuing the continued existence of a variety of forms of suffering and limitation of human potential seems at a minimum bizarre and perhaps even pathological.

The inclusion of homosexuality on Somerville's list is striking. To say the least, having a homosexual orientation is far less obviously a disability than, say, lacking an important human sense such as hearing (I'll return to this issue in the appendix). Moreover, the continued presence of a significant percentage of homosexuals may well add complexity and color to social life, although that might turn out to be a transitory social phenomenon. I expect that I am not alone in valuing such complexity and color, but liberal societies must tolerate those who disagree with me. This makes it problematic to use political coercion such as criminal prohibitions to override their values and autonomy. On the other hand, there are many ways short of political coercion for individuals, private organizations, or the state to combat any tendency for enhancement technologies to reduce some valued kinds of diversity.

For its part, the state could allow, or even assist, gays to have access to such technologies as PGD and genetic engineering based on an assumption (which might be testable) that they would, on average, be more likely than others to make use of them. After all, gays would be more reliant on them for having children. They might also be more prepared to choose children with a genetic propensity for a particular sexual orientation, if this were not prohibited for other reasons.

None of this is to suggest that, all things considered, it is desirable for parents to choose such genetic potentials, if they exist, as the potential for a sexual orientation. Nor is it to deny that an ability to do so could have social repercussions that might, in turn, lead to harms to homosexuals. Debate over this possibility has produced a rich body of literature over several decades (see Murphy 2012). Perhaps actual homosexuals would suffer harms in a society where these practices were commonplace. That, however, is a different contention. At this stage, my point is a rather narrow one: that arguments based on the possible long-term emergence of evils—or dubious harms to society as a whole—are generally unconvincing.

Similar claims can be made about the *Brave New World* scenario along with the possible corruption of such institutions as sports and entertainment, or daily life. Undesirable outcomes are not impossible, or at least not obviously so, but once again they are speculative. At least some of them are, moreover, dubious in the sense that one person's corruption might be another person's valuable innovation. I have mentioned my skepticism that enhancement technologies will lead to anything like the brave new world, and I should also refer to Feinberg's discussion of such issues as the corruption of taste.

Feinberg intimates that he would not wish to see an outcome in which the majority are devoted to lotus eating, or the exclusive reading of pornography or similarly stereotyped and simplified forms of literature that are (as he imagines the situation) likely to give an unrealistic picture of the world as well as reduce the capacities for subtle feelings and judgments. But if this provides a case for prohibiting anything, it is *all* such forms of what he considers cynical pseudoliterature (Feinberg 1988, 54–55). This seems to prove too much, since the impact would go far beyond the prohibition of pornography to affect many kinds of literature that are generally considered innocuous. Indeed, it could have broad effects, embracing much of daily life (should we be allowed to indulge in the often-simple emotions inspired by spectator sports when we could be reading the works of Henry James, James Joyce, or Jane Austen?).

In principle, it is possible that some future use of enhancement technologies will lead us down a path to the dulling of our feelings and judgments, yet that is speculative. Might it not actually have the opposite effect, if we are able to choose genetic potentials for *more* subtle feeling and judgment? It is not just a case where we should be skeptical, as with the idea that the mass availability of pornography will drive out the appreciation of Saul Bellow. Beyond such skepticism, we have reason to think that *more* people might be capable of appreciating Bellow's novels—and motivated to make the attempt—in a world with relatively little impediment to enhancement technologies.

Given all these vagaries, I henceforth will restrict my exploration of unwanted outcomes to those along the lines discussed by Devlin—threats of one kind or another to social harmony or stability. Here, what is feared is not some contentious and speculative evil, or some arguably regrettable outcome (as judged by a standard that may or may not be widely accepted). Rather, the claim is that social disintegration could take place, with clearly harmful results.

Devlin's Approach

Background

In the late 1950s and early 1960s, Devlin defended the use of criminal law to preserve and enforce the moral norms actually accepted by the general public. He argued that "immoral" practices, or "vice," in the relevant sense might properly be forbidden in order to protect society against what he called tangible and intangible harm (Devlin 1965, 111). He developed this position in response to the 1957 Wolfenden Report—commissioned by the British government from the Departmental Committee on Homosexual Offences and Prostitution.

At the time of Devlin's contributions, Mill's harm principle enjoyed considerable prestige with British law reformers and had influenced the conclusions of the Wolfenden Report. Devlin replied to the Wolfenden Report by defending the use of criminal law to suppress "immoral" conduct that lacked obvious victims, such as prostitution and consensual homosexual acts. He concluded that the harm principle was inadequate to determine the law's role with so-called victimless crimes. Such victimless conduct, he held, could cause both tangible and intangible harm to society as a whole.

Tangible Harm

Devlin (1965, 111–113) used the expression tangible harm to refer to a cumulative weakening of society if individuals weakened their capacities by debilitating indulgence in "vice," though he discussed this concept in an abstract way, and it is not clear that he believed homosexual conduct sufficiently debilitating to count. Note, though, that the concept of people making a social contribution—and possibly impairing their own ability to do so—is not entirely remote from Mill's thinking. Mill (1974, 70) viewed individual citizens as being obliged to contribute to the social good by, for example, assisting the national defense and the work of the courts. Thus, his perspective allowed for harms by omission, where an individual failed to meet these social responsibilities. He did not maintain that mere inactivity is always innocent or that law could not impose positive responsibilities.

That being so, it might seem possible to contend, without straying too far from the spirit of *On Liberty*, that liberal democracies can rightly demand that each citizen maintain at least some useful level of ability to enable them to meet their responsibilities. Yet Mill would have resisted such a conclusion. While he doubtless saw room to criticize shirkers, he recognized a role for the law only if an individual actually violated an obligation to others or at least created a definite risk that a specific obligation would be violated. Mill (ibid., 147–151) explicitly rejected punishing people merely for not taking proper care of themselves.

Modern liberal democracies need not accept every nuance of Mill's thought—it is not a holy gospel of timeless application—but there is some wisdom in his lenient approach to positive requirements to maintain our capacities. I take it that we do not want to be subjected to an overreaching public morality that imposes endless requirements on us to look after our own safety and develop whatever capacities our society might need, or favor, from time to time. Nonetheless, once we accept that we have positive obligations to assist others and contribute to some social objectives, it is difficult to draw an obvious, principled line. There is doubtless also some conduct that is sufficiently debilitating to reduce our ability to meet the sorts of obligations listed by Mill.

It follows in principle that society as a whole could have some legitimate interest in the talents, capacities, and skills of its citizens. But at the same time, it is rational for anyone who values individual autonomy to fear the ramifications if the state became too enthusiastic about making policy in this area.

In any case, none of this supports sweeping political coercion to suppress the use of enhancement technologies. It may instead tend to justify the use of moral or legal pressure to *encourage* the use of PGD, genetic engineering, and other methods that might even increase the potential abilities of its citizens. John Rawls saw the abilities of citizens, at least in part, as a social asset for common advantage. Accordingly, he argued in *A Theory of Justice* that society should preserve the general level of genetic potentialities and "prevent the diffusion of serious defects" (Rawls 1971, 107–108; 1999b, 92).[3] Rawls's overall theory of distributive justice may be extreme in the degree to which it supports societies making demands on the talents of their individual citizens, but perhaps it is *only* a matter of degree. Actual societies do, of course, place great importance on their available pools of developed talents, which partly explains the prominence of education and training as items of public expenditure.

After all, at least one reason to value our developed abilities is that they enable us to act so as to benefit those around us, contribute to our society as a whole, and assist other human and nonhuman beings whose welfare we care about. We should in that context once again note the synergies and network effects that can arise as increasingly greater numbers of individuals have access to enhancements (Buchanan 2011b, 48–50).

This might be another reason to oppose detrimental forms of genetic engineering, but it will generally be an argument *in favor of*—not against—such things as the use of PGD to choose embryos with an above-average potential for, say, intelligence or stamina. Similar considerations apply where genetic engineering has been used to boost genetic potential beyond what it would otherwise be. Since these interventions are aimed at magnifying rather than diminishing individuals' abilities, they would, if successful, tend to strengthen the talent pool and hence collective capacity of the society concerned.

Consider the example of Abigail and Belinda, explored above in chapter 3. As a reminder, Abigail has effective control of an early embryo that she intends to have implanted in her own uterus, but only after its genetic potential has been altered for greater muscular strength, expected longevity, intelligence, and resistance to disease. When Belinda, the resulting child, eventually comes into existence, it is rational to think of her as having received a benefit rather than as having been harmed. Beyond this, however, no tangible harm (in Devlin's sense) has been suffered by the society where mother and daughter live. Certainly, there are possible scenarios in which Belinda's additional intelligence and other characteristics might prove detrimental to her society; *you never know*. There are also possible scenarios in which her society might end up better off if she suffers from a minor or major disability. But that cannot support a judgment that Abigail's actions should be forbidden or even discouraged in some way.

Even less beneficial uses of enhancement technologies do not necessarily involve any tangible harm of the kind described by Devlin. Consider, for instance, reproductive cloning, the use of PGD to select the sex of a planned child, or the use of PGD or genetic engineering to choose some trivial characteristic such as eye color. Assume for the sake of argument that these actions can be carried out safely. On that assumption, they appear to be more or less neutral in their effects on citizens' abilities to carry out social responsibilities of the kind recognized by Mill or make any other wider social contribution. There could be a negative effect in some restricted circumstances; for example, some parents might deliberately choose embryos with the potential for a disability, low intelligence, or physical weakness. Yet these are likely to be uncommon parental actions.

A more plausible concern is the possibility that some children might be harmed psychologically by knowledge of their parents' actions—this is similar to a concern raised in chapter 4, and it must be taken into account out of solicitude for the welfare of children, irrespective of the harm to society's future talent pool. Generally speaking, though, the conclusion must be that considerations relating to tangible harm do not provide the materials for good arguments against enhancement technologies.

Intangible Harm

It was Devlin's concept of intangible harm that attracted the most interest during the debate that followed the Wolfenden Report. This notion has more promise than that of tangible harm as a theoretical justification for the prohibition of enhancement technologies or some of their possible uses. It also has the potential to pose great dangers to individual liberty and autonomy.

By the expression intangible harm, Devlin (1965, 113–115) means harm to a society's (supposedly) shared fabric of moral norms. To elaborate briefly on this, Devlin's writings maintain that every society has a fabric of ideas of right and wrong that is essential to it, but can be damaged by the proliferation of moral divergence. If toleration is extended to some acts that are contrary to these ideas, it can harm the entire fabric. That, in turn, can threaten the viability of the society itself (ibid., 9–15, 89, 114–116). Unless we are considering a highly repressive regime, such as that of Nazi Germany or South Africa under apartheid, this would be a bad outcome by almost anyone's standards. Even in a case like that, some social continuity might be preferable to a complete and sudden disintegration into a Hobbesian war of all against all. Devlin's conclusion is that a society is justified in enacting criminal prohibitions against practices that are contrary to its fabric of moral norms.

Before going any further, I should observe how implausible all this seems as an approach to settling debates about the legal regulation of sexual conduct, which was the context of Devlin's writing.

Intangible Harm and Sexual Mores

In contemporary Western societies, many moral norms are *not* shared by all individuals, cultures, and subcultures. Many practices and ways of life are socially tolerated, even by those who disapprove of them. Since Devlin's time, this has become even clearer with respect to consensual sexual practices. It thus is difficult to identify sexual practices and orientations that could actually cause intangible harm in Devlin's sense, without also causing more obvious harms to the agent or other individuals.

If we could identify a sexual practice that does not cause obvious harms yet is nonetheless disdained by many people in the society concerned, it is still not apparent how its legal toleration could be expected to precipitate social disintegration. This is partly because the citizens of liberal democracies have no reasonable expectation that the law will enforce a single, shared, extensive set of moral norms. As Douglas Husak (1987, 230) points out, to the extent that such a society has a common morality that provides its "cement," this morality involves "mutual tolerance and respect." The burden of political justification seems to fall quite opposite to the way proposed by Devlin. If it is argued that a specific practice should be prohibited because its toleration would cause social instability or something similar, cogent evidence and rationales will need to be adduced. A case will need to be made as to just what wider harm might be done to important moral norms, and by what psychological or other mechanism.

Indeed, our experience with social pluralism might lead to skepticism about what harm would be caused by the toleration of any specific practice *even in less liberal and pluralistic societies*. Imagine that the citizens of Kazanistan, a hierarchical, although relatively moderate, Muslim nation, have a shared moral disdain for consensual homosexual acts.[4] Should such acts be banned in Kazanistan because they cause intangible harm? At least in principle, empirical evidence might be available to support a claim that legal toleration would lead to the destruction of crucial, widely shared moral norms going well beyond those relating specifically to homosexuality. This might seem more likely in Kazanistan than in the societies of the West, since the former would no doubt have less experience in tolerating a diversity of moral norms.

Still, some of the reasons to doubt the likelihood of social breakdown apply even to Kazanistan—just as they do to New Zealand, Canada, or Sweden. Recall that the principle of intangible harm is ultimately intended to protect the harmonious or stable functioning of societies. Accordingly, the principle would seem to justify the legal enforcement only of those moral norms that meet two criteria:

Criterion 1 The norm is so structurally important to the relevant society's total fabric of moral norms that a widespread failure to abide by it will unravel the total fabric, and consequently, the stable and harmonious functioning of the society concerned.

Criterion 2 The norm is actually likely to be widely flouted if its enforcement is left to individual conscience or social censure.

If the second criterion is not met, then the intangible harm is not likely to eventuate, even in the absence of legal prohibitions. If the first is not met, the supposed harm will not eventuate in any event. Can a moral norm relating to sexual practices, such as the belief or sentiment that consensual homosexual acts are wrong, simultaneously meet both of the criteria that I have proposed? It seems unlikely in practice. A moral norm that condemns consensual homosexual acts might well meet criterion 2, such that it would be widely flouted without legal prohibitions, but there is little prospect that it could meet criterion 1, at least in the majority of actual societies.

In support of this, a moral norm according to which homosexual acts are impermissible can easily be distinguished from other norms that relate to such values as nonviolence or honesty. The latter are independently attractive, since they connect closely with fears of direct harms to our interests. Indeed, the operation of such norms appears necessary for the survival of any society at all, even if there is variation in the precise form that they take in different societies.[5] To elaborate slightly, it is unsurprising that social condemnation is commonly displayed when individuals obtain advantages by engaging in violence, lies, or deceptions, or flouting the local regime of property allocation (even if that allocation appears unjust when viewed from a broader philosophical perspective) (Hart 1994, 200–201). Nor is it surprising that most legal systems identify many kinds of violence, deception, fraud, and economic misappropriation as serious crimes. This pattern is far too robust to be threatened by the toleration of such "vices" as homosexual conduct.

It might be asserted that there is a middle ground between those moral norms that are almost inevitable and are necessary for the functioning of any society at all, and those that are somehow unimportant or inessential. The midway position might be moral norms that are important to *this* group's way of life, though not essential for absolutely *any* kind of social functioning among human beings (Feinberg 1988, 45–47). Perhaps, it might be claimed, the law should at least enforce these intermediate moral norms, possibly to preserve the society's existing way of life for its own sake, or possibly because the society's citizens will lack the resources to distinguish between intermediate norms and truly universal ones: if the intermediate norms are not enforced, the entire fabric will unravel, and disorder will result.

The issue may be arguable in respect to a society such as Kazanistan, though many of the countervailing considerations apply even there. In particular, no amount of leniency in allowing practices that have been condemned by tradition will remove Kazanistani citizens' good reasons for enforcing standards of nonviolence and honesty. Such standards are necessary for any kind of social survival and are unlikely to be abandoned even if they lose whatever other rationale they might be given by local traditions. In any event, consider the circumstances of liberal democracies, whose policies I am most interested in. In a country such as the United Kingdom, there was never any reason to believe that those of us who reject norms condemning homosexual acts would be more inclined than anyone else to be (for example) violent or dishonest. The same applies to homosexuals themselves: there was never a good reason to expect gays to be morally vicious in any identifiable way that transcends their sexuality (cf. Hart 1963, 51).

As events have unfolded since Devlin developed his principle of intangible harm, homosexuals have gained significant legal protections, many sodomy statutes have been repealed or struck down by the courts, and the cultural fabric of a typical large Western city now includes a prominent gay "scene."[6] The debate has now moved on to whether legal recognition should be given to same-sex marriages.[7] I am aware of no proof that the increasing legal and social acceptance of homosexual conduct in recent decades has actually made people in liberal democracies more violent or prone to dishonesty. Even if norms relating to homosexuality held an intermediate place in some traditional way of life in preliberal England, challenges to these norms along with a refusal to enforce them have not ushered in social disintegration. These traditional norms of conduct simply did not and do not stand and fall with norms relating to such problems as violence and dishonesty.

Similar arguments could be developed for most forms of consensual sexual conduct involving appropriately mature people. In all these cases, we should be suspicious of fearmongering about harms to a society's moral fabric that just *might*—by a remote and consequential process—lead to more obvious harms if someone let loose the hounds of sexual chaos.

Future Technology and Possible Harms

Intangible Harms to Society
Could the concept of intangible harm be more useful when applied to enhancement technologies? Not obviously.

Assume that moralized abhorrence of such innovations as human cloning and genetic engineering for enhanced capacities is just as common as was once the case with moralized abhorrence of homosexual acts. This is not altogether implausible. For example, Jonathan Baron (2006, 68) reports his research findings that many people flatly oppose reproductive cloning, irrespective of its benefits, and in a way that is unsurpassed by any other issue in its arousal of strong emotion and absolute moral commitment. Nonetheless, such opposition to reproductive cloning is not universal. Francis Fukuyama and Franco Furger (2006, 419–429), for instance, report data from a wide range of US surveys from 1993 to 2003, asking questions relevant to reproductive cloning. While the picture is mixed, and it is clear that opponents of reproductive cloning are the overwhelming majority, it is also apparent that not-insignificant minorities in many of the surveys find the practice (variously) morally acceptable, socially tolerable, or even personally desirable.

It would be fascinating if research could be gathered to show whether those of us who are prepared to tolerate, or even approve of, reproductive cloning are more violent or dishonest than our fellow citizens. As with tolerance or approval of homosexual acts, however, this is prima facie unlikely. Moral norms against violence and dishonesty are deeply grounded in their own right, and do not depend on their integration with any fabric of moral beliefs that includes a rejection of reproductive cloning (or other widely impugned practices such as sex selection, or genetic engineering for enhanced physical or cognitive potential). In short, if the principle of intangible harm is understood, as it was by Devlin, as relating to the damage to a moral fabric, leading consequentially to social disintegration, it has nothing to offer to current debates about human enhancement.

But intangible harm can be reinterpreted more broadly than harm to a fabric of moral norms. In Devlin's theory, the ultimate aim is evidently to protect social stability. The intention is to sustain the positive moral norms not for their own sake but instead only as a means to this end. We need not think of threats to social stability as consisting in or operating via the undermining of conventional moral norms. If the aim is to protect a society from threats of disharmony that could culminate in instability, collapse, or disintegration, we should focus on this threat directly, and not just on the particular means by which it might be realized. In principle, there might be *many* ways in which the cumulative effects of specific practices could be socially damaging.

This theoretical possibility is especially relevant if some practices have the potential to change the way human beings in the society naturally respond to each other. I have argued that core moral norms, such as those relating to honesty and nonviolence, are unlikely to be threatened by the toleration of such practices as consensual homosexual acts. Nonetheless—so it might be claimed—there are certain preconditions for the functioning of liberal democracies, and indeed there are some preconditions for the functioning of any society at all. Such things as the following might be essential to liberal democracies: the repudiation of violence and dishonesty as means of obtaining resources, or competing with others for social status and sexual success; the rule of law; equal legal rights for all; solicitude for the safety and welfare of children and young adults; freedom of belief, worship, and expression; and toleration of a wide range of value systems and ways of life.

This leaves a large area for individual liberty and social pluralism. People can be permitted to act in many ways that might ultimately lead to changes in common social practices and prevalent moral beliefs. Such changes are expected, and one of their recognized causes is the development and adoption of new technology. At the same time, the fabric of successful day-to-day social interaction involves the continual reciprocation of understanding, concern, and respect. This could be threatened if some future development began to undermine human beings' natural responsiveness to each other, or our sense of being involved in arrangements of reciprocal care and forbearance.

Much of our responsiveness is more or less unconscious and automatic. For example, we have an extraordinarily complex repertoire of facial expressions that has evidently evolved to express emotions; it is almost totally constant across human societies (Ekman 2003, 1–14). From New York to New Guinea, human beings are able to interpret essentially the same code of emotional signals, revealed by subtle movements of facial muscles. At the individual level, a human being who lacked the normal repertoire of emotional expression or typical psychological capacity to understand it would be severely disadvantaged.

Thus we would be harming a child, rather directly, if we altered her genetic potential so that as she grew to adulthood, her face and voice did not express emotions to us and elicit sympathetic responses from us, in the way of most human faces and voices. Perhaps we would give her intellectual recognition as an appropriate candidate for moral consideration, since we would be impressed by her status as a sentient, rational, and emotionally vulnerable being. Yet her moment-by-moment ability to participate in human society would be greatly hampered by what we had done. Unsurprisingly, that should be off-limits.

What if some technological advance undermined the responsiveness of human beings to each other in a more general way? This might even be done in such a manner that it is doubtful whether any identifiable individuals would, overall, be harmed. Consider someone who would possess great abilities, perhaps including a supple capacity for conveying emotions, while also being genetically and neurally modified to act unsympathetically to others. This might be the effect of a psychological disability in modeling human emotions or perhaps from a simple lack of care for others' sufferings. The diminution of her responsiveness might be an unwanted side effect of a genetic engineer's activity, perhaps resulting from the pleiotropic character of certain genes that must be modified to gain an intended benefit such as greater intelligence, longevity, or resistance to disease. On balance, this genetically engineered individual might actually be benefited rather than harmed. Yet for our own and each other's safety, we should avoid creating beings like this.

If affective communication or the emotional responsiveness of human beings to each other broke down in any widespread way, interpersonal interactions would be severely damaged and the functioning of the society concerned would come under threat. A simple example would be a scenario in which parents were able to engineer their children to have psychological traits that would benefit them in existing social circumstances—circumstances in which *others* are not so engineered—while being collectively damaging if the trait became common. In other words, there might be some traits that are valuable for individual success, but collectively disadvantageous if they become common within a group. This might apply, say, to some kinds of extreme competitiveness. It could be seen as the opposite of a beneficial network effect.

Any such scenario would eventually bring about tangible harms to individuals, though by indirect and consequential processes. Particular harms to individuals are speculative at the time that the technological intervention takes place. Even if the overall social impact is predictable, the harms to specific individuals—when the structure of their society begins to disintegrate—cannot be known in advance. All that can be said with confidence is that *if* the society disintegrates in some way, then the benefits of social living will be lost. If the outcome is a war of all against all, or merely the loss of social and technological infrastructure, it is not speculative to conclude that individuals will *then* suffer. In such an instance, life could be expected to grow nastier, more brutish and solitary, and shorter.

Imagine a different scenario: no changes have been made that are obviously harmful, at least on balance, to any particular individual. It is merely that a large gap has grown between the abilities of the society's enhanced and unenhanced humans, so much so that members of the two groups cease to identify with each other and each other's interests. Assuming that the normal repertoire of voice tones and facial expressions has not been tampered with, individuals from the respective groups might find that they still respond to each other with empathy and concern—*if* they ever interact in person. But they might, in fact, seldom meet, especially in circumstances conducive to any meaningful interaction. Enhanced and unenhanced citizens might inhabit such different social worlds, and have such different preoccupations and needs, that it no longer seems natural for them to take each other into account.

Such a society, with differential availability of genetic enhancement, might come to be dominated by a wealthy "genobility" with a concomitant genetic underclass (Mehlman and Botkin 1998, 98–102). Many people might have no prospect of improving their own lives, or at least those of their children; for them it would be a life without hope. Mehlman (2003, 115–116) suggests that it is only the belief that there is equality of opportunity, and with it the possibility of upward mobility, that makes current inequalities tolerable—but this belief would be eroded. In fact, the issue might no longer be one of *equality* of opportunity, but (for many) the continuing existence of any real opportunity at all. Apart from the effects on individuals, a society like this is one that could be threatened by serious tensions and even the possibility of collapse.

This conclusion certainly might be questioned, since previous societies have survived with inherited, hierarchical social status, but that might not be compatible with Western democracy as we have experienced it. One might imagine the emergence of various unattractive scenarios, such as more or less totalitarian rule by an overclass, or the provision of benefits to an underclass to keep it content (this brings back the otherwise-unlikely possibility of a brave new world). Alternatively, we could see political swings that create an unstable struggle between populists wanting to distribute enhancements more evenly while opponents attempt to maintain the privileges of the genobility. In any event, there might be significant instability and a risk of social collapse (Mehlman and Botkin 1998, 102–104).

As with the issues to be discussed in the next chapter, an obvious policy response would be the exercise of the state's great *economic* powers in an attempt to provide universal access to the same benefits, or at least to a reasonably approximate substitute for them. These might not be practical options, however, if the benefits were obtained by such a technology as genetic engineering, since universal provision could well prove expensive—beyond even the resources of the state.

Mehlman and Botkin argue strongly against the idea that costs would go down to make the technologies affordable or that the technologies would be self-funding. They also reject the notion that governments could find ways to extend universally all (but only) the nontrivial benefits from genetic technologies. Apart from issues of expense, there are difficulties with defining what is to be considered trivial, and there is no real precedent in extending such benefits, given that even public education does not extend the benefits of private education universally. Mehlman and Botkin (ibid., 105–108) doubt that society could adapt gradually to such changes; they assert (rather swiftly, I think) that ideas of gradual adaptation underestimate the profundity of the challenge to equal opportunity.

Possible Responses

It is not obvious that prohibition is the best response to these possibilities. In all such cases, the harms to individuals are relatively remote and even the social instability is caused indirectly. This leaves open the potential for more tolerant action to obviate the harm, such as by campaigns directed at lessening prejudice rather than preventing the birth of people who might become its victims (Harris 2007, 23–25). This is one answer to such apocalyptic views as those of George Annas, Lori Andrews, and Rosario Isasi (2002, 162, 173; Annas 2005, 36–38), who warn that differences arising from technological alterations to human capacities (not necessarily using the methods of genetic choice under scrutiny in this study) would lead by some indirect process to genocidal conflicts.

Under some conditions, however, the harm that might be done to society and ultimately individuals need not stem from mindless prejudice capable of being cured through reason and tolerance. For example, a change in the mix of personalities to include a much higher proportion of aggressive or extremely competitive people does not seem curable in such a way. Much the same can be said of the creation of people who are less psychologically equipped to offer and/or receive empathy from others; it is likely that social harmony requires a certain degree of prerational responsiveness between human beings, and that any erosion of this will not be curable merely by such steps as moral education.

These considerations might justify a degree of caution with some kinds of genetic engineering and perhaps even the use of PGD to select personality traits, if this became possible. Note, though, that many uses of enhancement technology could not reasonably be prohibited on any such basis. Many interventions would not even be plausible candidates for legal prohibition. It is not obvious that arguments based on intangible harm can be extended to support the prohibition of, say, human reproductive cloning, if it ever became feasible and safe. Nor can they be applied easily to embryonic sex selection (or sex selection by sperm sorting, if it comes to that). Reproductive cloning would merely repeat an existing individual genome in a new environment. There is no redesign involved, and no creation of beings less able to feel empathy or engage in affective communication. Accordingly, reproductive cloning would pose no clear threat to the psychological basis for our ability to live together in a mutually responsive manner.

Nor is any redesign involved in sex selection; there is merely a choice between two naturally occurring possibilities, either of which could have eventuated with no technological intervention. In principle, the widespread use of sex selection might cause disharmony in another way—via significant changes to sex ratios—but there is no reason to believe that this would actually come about in Western societies. On the contrary, survey data from several major Western nations show no strong preference for either sex (Fox 2007, 16). Perhaps there is a case for collecting and monitoring data from fertility clinics to ensure that no strong skewing of sex ratios is resulting, in which case a more liberal solution than total prohibition might involve regulation to require that sex selection be restricted to family balancing situations, where a couple who have several children, all of one sex, want a child of the other sex. At the moment, no case for such regulation has been made out.

It should be added that this kind of regulation would not *obviously* be needed even in a society in which there was widespread preference for one sex over the other. The claim that a significant change to sex ratios would produce disharmony is speculative. We might just as easily conjecture that it would do no more than create some social pressures toward greater acceptance of gay or lesbian relationships, or perhaps of polygyny or polyandry (depending on the direction of the skew). It is not at all clear that this would be disruptive if it happened gradually. Even if we assume that it would be, it is not apparent that merely allowing sex selection would create any significant *additional* skewing of sex ratios, given that less sophisticated methods such as ultrasound and abortion are already available.

What of the possibility of a massive divide—perhaps even the occurrence of something like a speciation event—between rich and poor? If the issue is not one of distributive justice, or fairness in some related sense, but instead of divisions that might reach the point of undermining social stability, consideration has to be given not only to the magnitude of the risk, *if it eventuates*, but also the probability that it will actually do so. Consideration should be given to policy responses that are less damaging to individual autonomy than outright prohibition of whichever innovations are seen as most likely to open up a damaging social divide.

Once the issue is viewed in this way, it enables the identification of a number of important aspects. First, it is not clear that those innovations that are most likely to be safe and practical will have the feared effect. There may be quite severe limits in general as to how far we could go with the creation of godlike masters, a genobility, or people who would seem like an alien species. Reproductive cloning would have no power to bring about those outcomes, and nor would PGD. The genetic engineering of human embryos is another matter, but even here we will be constrained, perhaps severely, in our ability to engineer viable people who would be notably different from us in either bodily morphology or psychological responsiveness and motivation (cf. Juengst 2009, 43–58).

Second, pace Mehlman and Botkin, it is not beyond the economic capacity of liberal societies to implement policies that could bring about greater equality of access to relevant enhancement technologies. Third, policymakers concerned about this particular problem have another alternative: rather than waiting for new technologies to appear, then trying to redistribute them in a way that reduces social divisions, they can confront the fact that many technological innovations have the potential to become differentially available to different economic classes, with a potential for different classes to drift away from each other. The long-term answer is not the piecemeal suppression of technologies that could have such effects but instead broader social reform to reduce economic inequality.

Finally, there is the possibility of introducing policies of amelioration. Mehlman and Botkin (1998, 125–128), for instance, argue that the best approach to reducing possible social instability is to implement a government-run "genetic lottery," the details of which they describe. In this approach, individuals and families who could not afford to buy beneficial genetic products could take part voluntarily in a chance allocation of funds to be used on the products of their choice. The lottery would not be used to raise funds (a ticket would not cost anything) but rather purely as a means to give poorer classes a chance to obtain access to genetic enhancements. The lottery could be financed from general revenue or via a tax on the provision of genetic enhancements (at various possible points in the chain of distribution).

This concept has attractions, such as providing the poor with opportunities to be more competitive with the wealthy—in terms of their individual cognitive and physical capacities. At the same time, it is not a complete solution to the problems of economic fairness, and it would not necessarily eliminate or even reduce feelings of resentment should the

wealthier classes gain access to significantly beneficial technological innovations denied to the masses or an economic underclass. I do not endorse
it as the preferred approach to public policy on enhancement technologies
in liberal societies, yet it provides one creative response and might play
a role. Regardless, we are not necessarily forced to adopt an illiberal approach built around the prohibition of safe and beneficial innovations.

Deliberations to produce the best policy mix will be affected by numerous imponderables. These include the timing of various innovations (how
soon they might become available and the order in which they appear),
the perceived benefits to society as a whole from the efforts of enhanced
individuals and the network effects of widespread enhancements, and the
expense actually involved. It will be necessary to find a policy mix that
is likely to reduce concerns about fairness and stability, without losing
the benefits of enhancement technologies, or trampling on the values of
liberty and autonomy.

The Danger of an Expanded Intangible Harm Principle

If we are to adopt an expanded principle of intangible harm, addressing
harms to society as a whole beyond those alleged to arise from challenges
to its moral fabric, then a key question to be asked of any innovation is
whether it creates a realistic threat of serious social breakdown or disharmony. This might occur, for example, if the innovation tends to threaten
the psychological bases of our ability to live together socially in a mutually responsive way.

Two important points need to be emphasized, however, before it is assumed that arguments about intangible harm to society actually should
be deployed to support a wide range of legal prohibitions. The effect of
these is to create a Scylla and Charybdis situation, though not something
as strong as a logical dilemma. The first point has already been explored
in the previous section: such arguments may not extend widely at all in
their application to contemporary anxieties about enhancement technologies. For instance, it is difficult to see how they could be used to justify
the prohibition of reproductive cloning or embryonic sex selection. Second, by contrast, on another interpretation the arguments may extend
too widely.

Some opponents of reproductive cloning, for example, might protest
that it would have a negative effect on social harmony, perhaps because
children who came into existence as a result of this technology would be
subjected to prejudice. This shows how the application of an intangible
harm principle could end up being far too wide, if not reined in somehow.

An extended principle of intangible harm, such as I have described, could become illiberal, indeed: a vast range of practices and social developments could have at least *some* tendency to damage social harmony, even if they also provided individual and social benefits. If the test is simply whether or not certain actions could tend to damage social harmony to some unknown extent, the legitimate operation of an expanded principle of intangible harm is almost unlimited.

Mill's *On Liberty* makes a similar point about the potential for overextending the harm principle once it is applied to anything beyond direct harms. Mill (1974, 158) fears a "monstrous principle" that anyone could demand as a "social right" that nobody else act in any way falling short of their own standards of perfection. Consider the potential for intolerant individuals to display impatient, unsympathetic, and even bigoted attitudes to others for an endless range of reasons. Almost any individual or cultural difference among the citizens of any society is a possible source of at least some disharmony. The underlying reasons can range from blatant racism to disagreement with others' religious, political, or metaphysical beliefs, to disapproval of their sexuality or distaste for their fashion sense. So we should be wary about accepting an excessively strong principle of intangible harm, such as the idea that *any* action at all that can tend to undermine social harmony is a legitimate candidate for legal prohibition.

If we accepted such a principle, it could have far-reaching and undesirable ramifications in many areas. At a relatively trivial level, we could find ourselves supporting prohibitions on personal choices such as to use Botox (Botulinum toxin) to resist the development of frown lines—after all, the cumulative use of Botox can reduce facial expressiveness and (at least to a small degree) harm affective communication. Many uses of the Internet might come under regulatory scrutiny, as the online world is notorious for incivility, flame wars, and the like. More dramatically, an excessively strong principle of intangible harm might be used to justify the dismantling of multiculturalism, the development of xenophobic immigration policies, and attempts to create a stifling monoculture. Such a principle would be far *too* strong to guide policy formulation in a liberal society, and would destroy the ideal of liberal tolerance along with the acceptance of social and ethical pluralism.

To underscore the point, we need only remind ourselves that liberal democracies show a spectacular, if imperfect, degree of mutual toleration of differences among their citizens. In such societies, many kinds of people can coexist and even flourish in reasonable harmony. Think of the differences that exist among the inhabitants of any typical Western city. Some are differences of preferred activities, some of beliefs, and some simply of personal characteristics. Here is an indicative (certainly not complete) list: the varied phenotypic traits, relating to body shape, skin color, and facial features, that are historically identified with "race"; minority sexual preferences such as lesbianism, male homosexuality, and bisexuality; greatly varying cultural practices and linguistic differences among various ethnic or national groups; dramatic alterations of physical appearance using such methods as tattooing, body piercing, and hairstyling; similarly dramatic impacts for personal presentation achieved through choices of clothing and accessories; deficits in organic functioning that affect perception and communication; and commitments to radically opposed systems of religious or metaphysical belief (some of which even portray unbelievers as villains in a cosmic drama wherein good struggles to overcome evil).

Much could be said about each of these as well as other differences among the citizens of liberal democracies. Some of the listed differences are, quite reasonably, regarded as misfortunes for the individuals concerned. Consider severe perceptual disabilities like blindness or deafness. These impose significant restrictions on an individual's access to knowledge of their environment, impede affective communication with others (e.g., a blind person cannot "read" facial expressions, and a deaf person cannot hear voices, much less discern tones of voice), and narrow the means by which the individual can take part in the larger culture (the visual arts are denied to someone who is blind, and the world of music is denied to someone who is deaf) (Glover 2006, 14–17, 23). Much can be done to accommodate blind or deaf individuals, but their disabilities are real.

By contrast, many other differences are not seen as misfortunes at all, and their presence is actually welcomed and even celebrated. Modern societies certainly would be relatively drab places if restrictions were placed on individuals' personal forms of presentation and style, even if these sometimes create communication barriers and could have some tendency, in some circumstances, to reduce social harmony. Despite that possible tendency, the presence in any society of different languages, cultural backgrounds, scenes, fashions, tastes, personal styles, and ideas is commonly experienced as enriching as opposed to a matter of regret.

A Better Intangible Harm Principle?

I suggest that we attempt to give some structure and specificity to an expanded principle of intangible harm. To begin with, there seem to be three problems with adopting the principle in a strong form. First, it would be counterproductive. Although many differences among individuals have some potential to undermine social harmony, attempts to suppress differences are likely to encounter resentment and resistance. The net effect is that it is better to allow a variety of practices the opportunity to flourish side by side. Second, we might be led to suppress much diversity that seems positively beneficial. Third, and perhaps most important, acceptance of such a principle would imply an undue pessimism about the practical ability of modern societies to accommodate difference by adopting an attitude of liberal tolerance.

In discussing Devlin's original account of intangible harm, I proposed that the burden should fall on those who seek to prohibit particular actions. They should meet two criteria, and any claims to do so in specific cases should be scrutinized with skepticism. The first criterion is essentially that the actions concerned violate a moral norm that is so structurally critical to the relevant society's total moral fabric that a widespread failure to abide by it will harm the fabric as a whole, thereby endangering social harmony. Second, the relevant norm actually is likely to be widely flouted if its enforcement is left to individual conscience or social censure (as opposed to prohibition).

Could similar criteria be used to restrict the operation of an expanded principle of intangible harm? On initial inspection, the answer is yes. The objections to an overly strong formulation would be met if an expanded principle of intangible harm were applied only to those practices that would become widespread if not suppressed by legal restraints, *and* would actually undermine social harmony (in some serious way) if they became widespread. Once again, we could stipulate that claims that some kind of practice or activity meets both criteria must be subjected to skeptical scrutiny.

Nevertheless, some further reflection is called for. Consider again the three problems for an excessively strong principle of intangible harm. One powerful reason for the state to decline the temptation to suppress particular activities is that suppression will provoke resentment and resistance. There is a potential for this to undermine the legitimacy of the state or its laws in the eyes of individuals who would otherwise be peaceful, honest, productive citizens. In such circumstances, attempts to apply a principle of intangible harm are likely to be divisive and counterproductive.

Such considerations apply to many areas of life, including expressions of sexuality and beliefs about religious matters. In these areas, important and personal interests are at stake (Hart 1963, 21–22). As Will Kymlicka (2002, 144) puts it in examining civil and political liberties (though he is not talking here about sexuality), they are crucial because "they allow us to have control over the central projects in our lives." Any attempt by the state to control adults' consensual sexual activities or practices (or refusal) of worship runs up against interests so important, and motivations so powerful, that they may override individual citizens' commitments to abide by their society's laws and conventions. Individuals' sexualities and their fundamental understandings of the world are likely to be experienced as given to or recognized by them, rather than as matters of choice. Attempts to control these areas of life are likely to cause deep suffering and provoke disaffection, if not outright rebellion.

By contrast, many aspects of personal taste and style may be less resistant to political suppression, but doubtlessly it would be resented if the state attempted to introduce new restrictions on the ways in which individuals might express themselves in (for example) their hairstyles, clothing, and manner of speech. Any new restrictions are likely to be experienced as tyrannical.

The situation with enhancement technologies is somewhat different. Technological control of such things as the genetic potential of one's children might become a jealously guarded freedom if we already had much experience of it, just as the freedom to educate ourselves and our children is something precious to us. Yet we have little experience to date of the use of genetic technology to bring about children through SCNT, choose characteristics such as sex or genetic potentials, or enhance potentials. It is more probable that there would be resistance to attempts to *take this away*, if it were already well established, than to efforts to prevent it from becoming available in the first place.

There nonetheless is the prospect that some uses of enhancement technologies will enable us to achieve deeply desired outcomes, such as children who could not come into existence without these technologies, children of a particular sex, or children with longer and healthier lives. To the degree that this is known, attempts to suppress such technologies, or prevent their development, will cause resentment and resistance that will be quite rational as well as reasonable.

Second, much of great value might be lost if the state and its agencies adopted a rigorous policy of suppressing anything at all that could undermine social harmony. Admittedly, there are many disagreements about what is valuable, and especially whether diversity should be valued in and of itself. Mill, for one, did not value diversity for its own sake but instead saw it as the outcome of arrangements in which people are free to develop their individuality. But he also thought that achieving this sort of freedom would require the creation of new practices that most individuals would then choose from, since creating them from the beginning is beyond most of our powers (Mill 1974, 129–132; Ten 1980, 71). Even if diversity is valued for its own sake rather than as a means to greater autonomy in shaping oneself, I have already questioned whether we should value every kind of diversity. There is, as Ronald Dworkin (2000, 441) reminds us, "no value, aesthetic or otherwise, in the fact that some people are doomed to a disfigured and short life." Similarly, it is far from clear that we should value the continued presence of a certain proportion of people with Down syndrome, as Somerville apparently does.

We need not commit ourselves to seeking more people in the world whose prospects in life are diminished by disabilities or low intelligence. Looked at from Mill's perspective, this kind of diversity does not empower us, as a diversity of available *practices* does, to develop our individuality. Harris (1998, 209) makes the important point that curing genetic defects will not reduce human diversity in any obviously *bad* way; for example, diversity of appearances would not be greatly affected while diversity of what people can *do* would actually be increased. Elsewhere he draws an analogy with literacy: the universality of literacy in large parts of the world has not diminished freedom and individuality; quite the opposite (Harris 2007, 128–129). The converse observation applies in the Abigail-Belinda illustration of engineering a human embryo for increased cognitive and other potentials. Other things being equal, Belinda's improved capacities would *increase* the variety of things that she would be able to do. If more people had enhanced capacities, they would be alike in that respect, but their actual life trajectories might vary more than those of "normal" folk without the enhancements.

Apart from the diversity of disabilities or illnesses, many other kinds of diversity might be criticized from one stance or another. From a vantage point hostile to religious belief, for example, many worldviews might seem to be little more than widely shared delusions transmitted generation by generation through culture and upbringing (Dawkins 2006). Conversely, from some hard-line religious perspectives, the political freedom not to believe in the "true faith" is really a freedom to reject the most important truths about reality.

Yet whatever the case with particular examples, many of us take delight in the overall diversity of modern pluralistic societies. The many personal styles, experiences, and ways of life on offer may be valued individually, or for the opportunities that they provide to many different individuals to find ways of life and shaping themselves. For many of us, value may be discovered in experiencing the sheer variety we encounter and diverse options we find available (Blackford 2012, 72). This gives us further reason to be wary of any jurisprudential principle that supports the suppression of all possible sources of social discord.

Thus, the successful suppression of enhancement technologies, or some kinds of them, could indeed result in the suppression of differences and benefits that might otherwise eventuate and be valued by many rational, reasonable people. Human enhancement might give many individuals abilities that they would value greatly, while the presence of individuals who have been altered in various ways by technological interventions, possibly including the enhancement of their intellectual and creative powers, might add more color and richness to society.

Third, a strong principle of intangible harm is simply not necessary. Its adoption would imply an undue pessimism about the ability of liberal democracies to accommodate group and individual differences. Admittedly, some changes that actually modified significant aspects of human bodies and minds (and went beyond the superficial level of cosmetic surgery or body piercing) might have an unprecedented potential to undermine our responsiveness to each other. In fact, they might radically strike at the sense of commonality among human beings that allows so much *other* diversity to be tolerated or even welcomed.

Though human enhancement may bring benefits, its potential to threaten social harmony does seem to be greater, at least with some conceivable innovations, than is the case with many other kinds of differences among citizens. Furthermore, the inevitable lack of evidence (either way) as to how well modern societies could cope with human enhancement may tend to favor prohibition or hostile regulation. In many other cases, of course, it is reasonable to throw the burden of proof on those who favor suppression of a practice. That is less obviously reasonable with genetic-choice technologies, because directly applicable empirical evidence is unattainable. Since some genetic choices are potentially quite radical and it is not reasonable to demand precise empirical proof of their foreseeable social impact, opponents should not be required to meet such a strong standard of justification as in other situations where intangible harm to society is predicted.

All that conceded, the standard cannot be the mere ability to engage in frightening speculation. Perhaps little warning needs to be given against the creation of beings who would suffer and perhaps be driven to desperation, like the monster depicted in Mary Shelley's *Frankenstein*. As for the idea that social stability could be undermined by the creation of a hierarchical caste society, it seems superficially plausible but subject to a number of caveats, already discussed and due for further consideration in the next chapter. In particular, there are practical limits to how far we could go in the creation of a genobility or race of godlike masters.

We should, in short, be wary of excessive reliance on the idea of intangible harm. In some clear-cut cases, it may have appeal, but it also has its dangers. Specifically, it could erode the beneficial social pluralism that is characteristic of life in a liberal democracy.

Conclusion

We should not underestimate the success of liberal societies in accommodating differences. In principle, many of these could put social harmony under some strain, although the tendency has been to accommodate them in the (plausible) belief that greater strain would be produced by attempts at suppression. Though pluralism and toleration doubtless have some limits, we would be justified in taking a skeptical attitude to any specific claim that particular innovations will tend toward social breakdown.

This suggests that arguments about intangible harm to society, analogous to those developed by Devlin, can provide only a limited basis for resistance to new social and technological developments. Great care should be taken before we stigmatize new practices as being likely to endanger social harmony. This warning has a wide application, but it applies with specific force to opponents of enhancement technologies.

7

Fairness, Equality, and Distributive Justice

Opponents frequently claim that enhancement technologies will undermine distributive justice, exacerbate existing inequalities, challenge ideas of socioeconomic equality, or subvert egalitarian political aspirations. Pence (2000, 70–71) observes that "the fear of biologically-based inequality has been one of the most substantial and pervasive objections to genetic enhancement (and also to human cloning)."

Many issues are entangled here, however, and in the thick of debate about equality and justice it can be difficult to establish just what argument is being put forward, and why it might provide a reason to prohibit some or all uses of technologies that offer genetic choice. Having raised the specter of social instability in the previous chapter, I now wish to consider contentions that involve more abstract ideas related to justice and equality. But there is some artificiality to this distinction. The issues are clearly related, since *one* reason why we might support relatively egalitarian social arrangements is that they are thought likely to encourage social stability.

Despite the potential for confusion, it is possible to distinguish at least three sets of concerns that relate human enhancement to fears of biologically based inequality. For one, there is the worry that combining existing wealth differentials with the availability of genetic choice would cause social instability. I examined this possibility in chapter 6. In this chapter, I will focus on two other sets of concerns.

One relates to the likelihood that existing social and economic inequalities will skew access to the benefits of enhancement technologies. Here, the issues relate to unfairness in the distribution of technological benefits. The other concerns involve more remote consequences: it is feared that the availability of genetic choices might create intolerable inequality in the future. The resulting social arrangements might be deeply unfair, or somehow oppressive to subordinated individuals or groups, even if the arrangements are stable and involve no obvious threat to the society's ongoing viability.

Arguments from Fairness

Walter Glannon is a clear example of the kind of critic of (some) enhancement technologies who concentrates primarily on issues related to fairness or distributive justice. His argument begins with the plausible claim that the genetic modification of human traits would in practice give some people an advantage over others in "beauty, sociability, and intelligence." Individuals whose parents could afford it would be granted an advantage over others in competing for social status, career success, and income, though the relative disadvantage suffered by others would not be caused by any fault of their own (Glannon 2001, 97–98). Note that Glannon's references to sociability and beauty show that he is worried about a wide range of traits that might boost an individual's life prospects.

For Glannon (ibid., 98–101), this argument can stand alone as a reason for prohibiting the use of genetic engineering for beneficial characteristics, even though he also advances other claims relating to autonomy, resentment, social instability, and the consequences of what I've called a Red Queen race. In developing the argument he maintains that there need not be any diminishing marginal value in increasingly high levels of human capacities (or other beneficial characteristics), so the potential advantages to be obtained by the rich would be unlimited. This may be an exaggeration, but it is a thought worth holding on to and considering specifically, and thus I'll return to it near the end of the current chapter. Meanwhile, it is worth remarking how Glannon emphasizes the *unfairness* of boosting the beneficial traits of some children, when no child deserves to receive the boost (or miss out on it) more than any other.

Even if enhancements to human capacities were universally available, Glannon (ibid., 99) points out, there would not be complete equality, since some parents would still be more financially able, or simply more willing, than others to take advantage of opportunities for their children to acquire the kind of education that would develop and utilize their capacities. This confirms that his concern is the possibility of some children gaining undeserved competitive advantages or suffering undeserved disadvantages, whether or not the disparity is caused by differential wealth.

Glannon is far from being alone in developing arguments along these lines. To take just one other example, Mehlman makes similar observations, although he is thinking of a wider range of technologies than the genetic engineering of embryos, and his primary stress is on what he sees as a threat to social stability and democratic arrangements. In the relevant passage of his book *Wondergenes*, he suggests that genetically engineered individuals will have superior abilities, and so will be advantaged in many competitive or partly competitive interactions—hence they may come to be regarded as "cheaters" (Mehlman 2003, 109–113). He then argues that their superior traits will be obtained through money that is itself either ill gotten or in any event undeserved because it can be traced back to other things that are undeserved, such as somebody's natural talents.

At this point in his case, Mehlman's argument is that enhancements to individuals' capacities and other traits will be deserved no more than are natural talents from chance combinations of genes. Possession of technologically enhanced traits would be seen as unjust—rightly so, in Mehlman's view—and *therefore* lead to resentment and resistance.

Arguments such as these depend on the claim that genetically engineered children with enhanced traits will have obtained a benefit—perhaps including a competitive advantage in gaining further benefits—that they did not deserve. These can be thought of as arguments from fairness. Arguments from fairness possess more initial plausibility when we consider genetically engineered children than when we consider children chosen by PGD or produced by reproductive cloning. What is special about the first category is that the very same children *could* have been brought into existence without the advantage to their life prospects. For the sake of simplicity, I will sideline issues relating to PGD and cloning. I also will tend to skip over the possible benefits to *parents* and instead will concentrate on those obtained by children if efforts are made to enhance their genetic potentials. This is where the alleged unfairness seems most plausible, since no embryo can have done anything to deserve this special treatment (or be denied it, if it is available to others).

A Scylla and Charybdis Problem

As a first impression, arguments from fairness may prove too much. On the one hand, *many* things are disproportionately available to the rich—think of expensive houses, fine wine, good grooming, fashionable clothes, fancy cars, piano lessons, international travel, and opportunities to meet famous or beautiful people. Their children are likely to receive such benefits as superior nutrition; private schooling; exposure to a broad range of art, literature, and culture; access to contacts in business, the arts, the professions, and/or government; socially admired grooming and clothing; opportunities for privacy and reflection; and the parental advancement of money, shares, and other income or property. Many of these things are highly advantageous, sometimes in subtle ways, to those who obtain them. They can have dramatic effects on life prospects.

Liberal democracies do not prohibit these things. In practice, they allow wealth differentials to develop, and do not attempt to eliminate private schools or the wide range of other advantages that the wealthy are able to provide for their children. Rather than preventing all advantages that are judged to be undeserved, public policy is usually confined to more modest social goals, like ameliorating the harsh outcomes that could arise from unrestrained industrial capitalism. Why should it be different when policy consideration is given to enhancement technologies? Why not permit the rich to use them for such purposes as to bring children into the world who could not be created in any other way, select children with particular traits, or enhance their children's beneficial traits? Given our current practices with private schools and many other goods that parents provide to children, where, exactly, is the problem?

If Glannon's standard of fairness rules out private schooling and the other advantages I've described—as well as certain uses of enhancement technologies—then it seems implausible. Or at the least, it is a standard that is far stricter than liberal democracies have been prepared to apply to public policy across the board. On the other hand, if his proposed standard of fairness does *not* rule out a wide range of advantages, how can we be sure that it will rule out exactly what he suggests? How can we be certain that it really will exclude genetic engineering to enhance beneficial characteristics?

Right from the start, then, the kind of argument pursued by Glannon runs into difficulties. Glannon needs a theory of justice and desert that indeed does rule out what he argues for, without undermining its own plausibility by ruling out much more. Unfortunately, he offers no theoretical account of distributive justice and desert, much less one that is at all adequate.

A Rationale in Political Philosophy?

With their emphasis on undeserved advantages and disadvantages, arguments such as Glannon's and Mehlman's evoke an idea that is popular in contemporary political philosophy. Rawls and many other liberal egalitarians have relied on the claim that at least prima facie, individual differences in natural talents (and hence the actual abilities developed from them), and in other advantageous personal traits such as beauty, should not influence the distribution of social and economic benefits, such as wealth, income, status, and political power. Nor should individual differences in family background and social class.

Presumably the benefits that are relevant here need to be given *some* limitation. To start with, what about moral or spiritual benefits such as the possession of useful virtues or knowledge of the methods (allegedly) required for spiritual salvation? The political philosophers concerned do not seem troubled about these, even though their possession by some people and not others was not deserved—at least not *ultimately* deserved if, as I've argued in chapter 4, no one is an ultimate self-creator (and so no one can be the *ultimate* cause of their own virtues, religious beliefs, and so on). But perhaps these specific benefits can be set aside because they are too controversial. Certainly no liberal thinker in the Millian tradition will require a fair distribution of purely spiritual benefits, since it is not possible to make any uncontroversial determination as to what the benefits really are or how, if at all, they are to be obtained.

Still, there may be other benefits—benefits in addition to spiritual and moral ones—that fall outside the purview of justice. In fact, there may be *many* benefits that should not be counted as "social goods" to be distributed fairly, or in accordance with a conception of justice. What about relationships, for example? Consider the case of two teenage boys attempting to attract the attentions of the same girl (or boy, for that matter). Common intuitive responses suggest that the outcome should have nothing to do with considerations of justice, and definitely should not involve any intervention by the state. Yet the outcome may greatly affect the respective individuals' life prospects. At the same time, it may well be influenced by the "winner's" qualities of charm, good looks, quick wit, athleticism, and so forth, which ultimately can be traced back to causes that lie beyond his control, such as his genetic potential and early upbringing. He did not deserve these things, but we are content for him to "get the girl" (or boy).

Before I leave the area of relationships, I should add that there is no clear line between "mere" relationships of love and those that involve economic concerns. We might stipulate that the two teenage boys in the previous illustration were not attracted to the girl because of her wealth or that of her parents. Wealth might well be a factor in other cases, however; it is easy to imagine situations where two potential lovers are attracted to the same woman or man partly because the object of their affections exudes an aura of wealth, sophistication, and perhaps power. The "prize" may well involve economic benefits and social status, but we do not expect the apparatus of the state to intervene to ensure that the distribution of wealthy lovers is, in some sense, fair.

What about parental relationships? Obviously, the children of wealthy—or merely willing and wise—parents do not possess anything like desert, merit, or prior entitlement to the advantages they receive from being born into a rich or otherwise-helpful family. Yet we normally think that the parents are (at least!) *permitted* to give their children various advantages, including advantages that will assist them later on in the (almost-inevitable) competition that they will face for social status, economic gain, desirable sexual partners, and so on. Such actions are not usually stigmatized as unjust.

Perhaps wealthy parents or wealthy potential lovers should not be allowed to obtain their wealth in the first place, though it is far from clear that they are prevented from doing so by any plausible theory of distributive justice, at least if they employed their developed capacities and skills to obtain wealth by lawful means. But once they have obtained it, why are parents not morally entitled to spend it in assisting the prospects of their children, including by such means as education, good nutrition, opportunities to make useful contacts ... and the prudent application of enhancement technologies? If we object only to the last of these, and not to the others, might we not be doing so only out of status quo bias, or some other disreputable fear or prejudice? *Perhaps* there is something special about enhancement. I'll return to this later in the chapter, but for the moment it's not apparent just what that "something" might be.

Nullifying Brute Luck

A Basis in Rawlsian Theory

For the sake of argument, let us assume that such benefits as genetically enhanced potentials are social goods that fall within the scope of a theory of distributive justice. At the same time, the genetic potential for advantageous traits cannot be deserved by somebody while she is still an embryo, so Glannon is correct that they are undeserved by the child who results, and that any relative disadvantage suffered by others is also undeserved.

The Rawlsian (and post-Rawlsian) tradition in liberal egalitarian philosophy relies on the idea that, as worded by Rawls (1971, 15; 1999b, 14) in *A Theory of Justice*, "the accidents of natural endowment and the contingencies of social circumstance" should not be used as "counters in a quest for political and economic advantage." That is because these "accidents" and "contingencies" are not deserved. As Rawls (1971, 74; 1999b, 64) puts it later, specifically with respect to natural talents, "the outcome of the natural lottery" is "arbitrary from a moral perspective."

Ultimately Rawls has far more to say—a fact that is often overlooked in claims such as those advanced by Glannon. Yet his starting point is that individuals should not benefit (in certain ways) from their natural talents because no one deserves these more than anyone else, and at the same time no one is morally entitled to (certain of) the fruits of being conceived or born into a rich, or otherwise-advantageous, family. Thus, if Belinda benefits from being Abigail's child, she receives a benefit that she did not deserve: she did nothing that she can use as justification for being the child of this particular parent. It is not something that was *due to her.*

Though Rawls eventually enunciates a principle that justifies some wealth differentials, one important part of his reasoning is the intuition that nobody should benefit or suffer relative disadvantage from initial circumstances in their life that they did not deserve (Kymlicka 2002, 127). Here I disagree with Buchanan and his coauthors (2000, 66–67) of *From Chance to Choice,* who see Rawls as (arguably) endorsing what they call the social structural view—a view that I discuss later in this chapter. It seems clear enough that Rawls is motivated not only by concerns about past acts of discrimination, exploitation, and so on, but also by his intuition that there is something problematic about "undeserved" natural talents and family circumstances influencing the distribution of social goods.

Following Kymlicka (2002, 127), I will refer to this idea as "the Rawlsian intuition."[1] Based on this idea, Rawls (1971, 86; 1999b, 100–101) states explicitly that undeserved inequalities of "birth and natural endowments" should be compensated for by some means that will "redress the bias of contingencies in the direction of equality." Therefore, at least in his starting point, Rawls is what Buchanan and his colleagues (2000, 67; following Thomas Scanlon) would classify as a brute luck theorist: someone who wishes to nullify the undeserved effects of luck. Rawls's famous difference principle, which allows some incentives for the contributions of the talented, is intended to do this to an extent (while providing a principled basis for the inequalities in social goods that remain).

Before I go on to consider other foundations for arguments from fairness, a series of questions arises. Should we share Rawls's hostility to brute luck and undeserved advantages? More specifically, should we accept the Rawlsian intuition as a basis for political philosophy? If so, how should it be carried over into public policy on such topics as human enhancement?

The Problem of Desert

Should we accept the approach that political action will nullify brute luck, at least where it affects the distribution of certain social goods? If not, then fairness arguments such as Glannon's must either be jettisoned or justified on some other basis. The supposed imperative to nullify the effect of brute luck, however, is far from uncontroversial within political philosophy. Many prominent philosophers have subjected this notion to extensive and severe critiques, including Robert Nozick (1974, 183–231), Joseph Raz (1986, 217–244), George Sher (1987, 22–36; 1997, 61–77), Harry Frankfurt (1989, 134–158), and John Kekes (1997, 88–136; 2003, 64–81). It is by no means obvious that the Rawlsian intuition will eventually prevail in political philosophy. Indeed, I submit that it should not.

This is not the place to review the full range of arguments developed by Nozick and others, which would require an extensive study in its own right. I should make it clear that I do not necessarily endorse any of their specific contentions except where explicitly stated. Nonetheless, I will explain at moderate length why I think that Nozick and the rest are on strong ground.

The arbitrariness, from a "moral point of view," of both family circumstances and "natural endowment" receives a strong emphasis in Rawls's work. But the idea of natural endowment in particular is highly problematic. I propose to distinguish, wherever necessary, between natural talents—using this to refer to something like genetic potentials—and a broader idea of natural assets that goes beyond raw potential, and includes such things as the capacities and skills that an individual has actually developed. It can, of course, be claimed that developed capacities and skills are *ultimately* undeserved, since they depend on (undeserved) natural talents, but this is not how we usually think about it. After all, many capacities and skills are obtained only after extensive training, education, self-discipline, and so on, and we might normally be tempted to say that somebody who now possesses these *deserves* to have them. In any event, we generally tend to think that there is nothing morally wrong with someone possessing them.

Since we are not ultimate self-creators, we could never *ultimately* deserve anything at all if desert were always dependent on its preconditions themselves being deserved. Reasoning such as Strawson's (2010; discussed above in chapter 4) establishes that everything about our own characters, talents, and developed skills can eventually be traced back to features of the world that we did not control, and hence did not deserve.

Since all our decisions in the end depend on our characters and principles of choice, and since these, too, are eventually traceable to events and states of affairs that we do not control, it can be claimed that we do not deserve what flows from them. This claim does not even depend on the truth of determinism. Arguably, the situation is not changed if the world is a mix of determinism and random quantum events (Mackie 1977, 226).

If all that is true, and unless Nozick is correct and the regress can be blocked, how can we ever deserve *anything*? Theories that attempt to nullify the impact of luck typically do so in order to distinguish between matters of luck that are not in someone's control and matters of personal choice, which (supposedly) are (Buchanan et al. 2000, 67). If, though, we were to go on looking for *ultimate* control, it would be difficult to sustain such a distinction.

To add to the puzzle, consider advantageous traits of character (in contrast to such things as cognitive, perceptual, and physical capacities, or skills associated with them). Our ordinary, pretheoretical notions of desert may support calling advantageous traits of character "deserved," particularly if the individual has deliberately cultivated them; they might well result from an individual's deliberate action or inaction (Sadurski 1985, 122). And yet even the decision to cultivate such abilities and dispositions falls within the general causal order. Such decisions—and the character traits themselves—ultimately depend on the individual's possession of a specific genotype and their exposure to a childhood environment that led to its being expressed in a particular way. If we nonetheless insist that such things as industry and self-discipline can be deserved, then we must agree with Nozick (1974, 225; emphasis in original) that the foundations for desert do not themselves—at least in all cases—need to be "deserved, *all the way down.*"

Perhaps it is best to observe that we do not, at least usually, deserve to *have* certain character traits, such as good or bad dispositions of character. They are largely the outcomes of circumstances beyond our control (genetic potential, early upbringing, etc.). Even if that is so, however, we simply might deserve praise or blame *for* them. On this approach, all that is needed when it comes to bad character traits, such as cruelty, is that their possession reflects badly on us. They can do this even if we did nothing to deserve to have them in the first place. I might also seem to deserve blame or punishment for a cruel *act* even though it happened as a result of dispositions that I did not choose. Admittedly, I may not deserve blame or punishment for an otherwise-bad act that was committed under duress or the disadvantage of a nonculpable mistake about the facts. But that is because the coercion or mistake negates any tendency for the act to reflect badly on me. If I commit the act without such an excuse, it certainly does reflect badly on me, and that perhaps is enough for me to deserve blame (Sher 2005, 51–70).

At this stage, we need a more comprehensive understanding of desert that stands as a good alternative to the idea that desert depends on ultimate self-creation. In the next (major) section of this chapter, I'll make a fresh start. First, there is more to say about Rawls's intuitive rejection of brute luck as an influence on the distribution of social goods.

Can the Rawlsian Intuition Be Rationally Grounded?

In *A Theory of Justice*, Rawls (1971, 17–22; 1999b, 15–19) describes a process of attempting to achieve equilibrium between the conditions specified in his thought experiment involving hypothetical contractors in the "original position" and what he takes to be our settled (abstract) convictions about justice. He seeks to obtain a fit between principles that can be derived from the original position and settled convictions that we supposedly have prior to deriving these, such as the conviction that success in obtaining (certain kinds of) benefits should not be sensitive to people's natural talents or family circumstances.

Still, when it is juxtaposed against many of the specific circumstances that can be adduced, the Rawlsian intuition fails to comport with common intuitive responses. For example, our intuitive response to a situation where two people compete for the same job is difficult to reconcile with the Rawlsian intuition. If each person attempts to prove that they are the best candidate, they will point to evidence of their developed skills, capacities, and work-related dispositions of character (they are hardworking, self-disciplined, etc.). Yet none of these is deserved in any ultimate sense, and they are all highly sensitive to the respective candidates' natural talents and upbringing. Such concrete instances might not matter if the Rawlsian intuition were the conclusion of an argument rather than the *beginning* of one. We might then adopt it *despite* its failure to match our intuitions about many everyday cases. Perhaps we could even develop a theory as to why those everyday intuitions so often go wrong (might they be products of false consciousness in some sense?).

It is difficult to imagine what uncontroversial premises could be relied on to reach such a conclusion. The Rawlsian intuition does not claim to be based on any empirical equality among human beings, for example, and nor could it be. As Max Hocutt (2000, 181) points out, experience shows that individual human beings are not equal in capacities, moral virtue, or anything else, and it is unclear why we should treat them all as equal when they manifestly are not.

The Original Position

It might be thought that Rawls does, despite all this, provide an argument that goes beyond the bare assertion that people should not benefit (in certain ways) from differences that are undeserved, such as differences in upbringing and natural talents. After all, he employs the whole panoply of his original position thought experiment to underpin his eventual account of distributive justice, which includes the difference principle. His most definitive statement of his principles of justice is in *Justice as Fairness*, where the difference principle is formulated as the principle that social and economic inequalities "are to be to the greatest benefit of the least advantaged" (Rawls 2001, 42–43). The principles of justice that Rawls settles on are supposedly those that would be agreed by hypothetical social contractors negotiating behind a veil of ignorance, including ignorance about their own identities, talents, and conceptions of a good life.[2]

Nevertheless, we are in danger here of getting things backward. Rawls (1971, 14–21; 1999b, 13–19) is totally candid that he does not use an argument about the original position to *support* his convictions about the undesirability of brute luck affecting distributions of social goods; rather, his specification of the original position *depends* on it. Thus Nozick (1974, 215) is correct when he complains that Rawls has designed the original position, with all its stipulations, to embody and realize his opposition to "allowing shares in holdings to be affected by natural assets." The original position does not provide an argument in support of the Rawlsian intuition but instead models certain values that come from elsewhere, including Rawls's conception of equality.

Accordingly, what emerges from the original position depends on what moral assumptions have already been made—or what values are assumed—when a description of the original position is specified. What emerges depends on what assumptions, intuitions, or values are being modeled by a particular description of the original position. It is clear, and Rawls certainly does not deny it, that the Rawlsian specification of the original position already *assumes* the truth of the Rawlsian intuition; the latter cannot be deduced from it.

To be fair, Rawls's thought experiment with the original position might still give some weak support to the intuition, given several assumptions. It might be proposed that *if* the Rawlsian intuition is prima facie plausible, *if* modeling it in the specification of the original position could (for all we know) have led to the choice of implausible principles of justice, and *if* we see that Rawlsian bargainers would actually adopt plausible ones, *then* the intuition passes a kind of (thought) experimental test. It might be said, then, that the prima facie plausibility of the Rawlsian intuition and the independent plausibility of Rawls's eventual principles of justice are mutually supporting.

Yet it is doubtful that the Rawlsian intuition has even this kind of weak support. The most obvious problem is that, as mentioned, it fails to conform to many common intuitive responses to everyday situations. This more direct challenge to the intuition's acceptability appears decisive. It is not as if we act and make judgments in ways that oppose the Rawlsian intuition only in marginal cases, while finding that the intuition explains a wide range of other actions and judgments. In fact, the actions and judgments of ordinary people run against the intuition clearly as well as frequently.

Given those circumstances, it appears almost irrelevant that the intuition can be modeled and employed in a thought experiment that might, after much complex reasoning, lead to independently plausible principles of justice. Compare a scientific hypothesis that accounts for some observations and survives some experimental tests—but is decisively refuted by other observations that are well known and rationally accepted. While scientists are generally reluctant to reject a highly successful theory because of a few anomalies, the lack of fit between the Rawlsian intuition and our intuitive response to everyday situations is more like the comprehensive failure of a new hypothesis to explain important observations. Under those conditions, whatever successful predictions the hypothesis makes will not justify its acceptance.

But even if this methodological point is waived, the situation remains murky. For a start, it is not clear what principles actually *would* be chosen by rationally self-interested contractors in the original position even if its features were those specified by Rawls. The contractors' bargain might be sensitive to many assumptions about issues that are (in the real world) highly controversial, such as assumptions about human nature. If, for example, the contractors believed that competition is inevitably an important part of life for human beings, perhaps because competitive urges are deeply ingrained in our nature, they *might* conclude that only a highly libertarian set of economic arrangements would be socially workable, at least in economically complex societies. They nonetheless might be sufficiently averse to risk to supplement free market arrangements with significant taxation in order to fund a generous economic safety net. Or they might draw similar policy conclusions from different assumptions—for one, they might reason that a principle of maximizing total wealth would be in their best interests as individuals, while again insuring themselves against individual disasters with an economic safety net.

Much more could be said about these difficult issues in modern political philosophy. At least two points are clear, though. First, Rawls's famous thought experiment does not provide strong support for the Rawlsian intuition by entailing its truth—and nor does Rawls make such an unlikely claim. On the contrary, the intuition comes first and the thought experiment then models it. Second, the thought experiment does not even give the Rawlsian intuition the kind of weak corroborative support that Rawls might reasonably hope for.

Desert: A Fresh Start

Our Acceptance of Competition

At this point, it is worth starting again, reflecting on our ordinary social experience to see whether we really do endorse anything that goes as far as the Rawlsian intuition. Such reflection quickly brings us to our everyday acceptance of a familiar fact that I have already foreshadowed. It is unremarkable, until it is specifically remarked on. During the course of our lives, we often encounter each other as economic, social, and sexual rivals. As we pursue a great variety of personal goals and ambitions, we frequently (not always, but certainly often) find ourselves in competition with others. We habitually draw on many talents and skills along with other characteristics or abilities—whether they are cognitive capacities, charm, pleasing looks, athletic ability, creative talent, or sheer doggedness—that we would not possess without a contribution of brute luck. We can be lucky or not so lucky with our genes, our places of birth, chance aspects of our uterine environments as we grow from zygote to birth, our family backgrounds and upbringings, and so on.

Take the familiar case, previously mentioned, of two or more individuals competing for the same job. Perhaps it's an attractive academic post at Oxford University. They and we might well accept that *some* competitive tactics are out of bounds. These proscribed tactics might include, for example, physical attacks on rival applicants, spreading false (or even true) rumors about rivals' embarrassing sexual habits, and stealing the rivals' outgoing mail to reduce the possibility of unwanted publications appearing in peer-reviewed journals. Yet with these sorts of constraints specified, the rivals' actions in applying for the job seem acceptable enough—that is, we do not object to the situation where these two or more people seek the same job in competition with each other. We certainly do not expect rival candidates to refrain from relying on, say, their records in teaching and research, even though their past successes in educating (and perhaps winning over) students, along with their various publication acceptances, will most certainly have been causally dependent on such things as their natural talents and other circumstances that they did not deserve.

Competition is a ubiquitous aspect of human life, expressed in the traditions of every culture. If we examine the question rigorously, going down the path marked out by brute luck theorists of distributive justice, we might eventually define all competitive advantages as unfair and embrace a moral system that rules out the entire idea of fair competition (Sher 1987, 76–77). But we do not adopt an attitude of blanket moral condemnation toward competition. It's not even that we start with an intuition that competition is morally problematic and then rationalize our way toward making a large class of exceptions on the basis of other considerations. Competition per se does not appear to be morally problematic in the first place, at least not if the everyday intuitions of ordinary people can be relied on.

It might be replied that some of the kinds of competition that I have briefly mentioned relate to things that fall outside the proper scope of distributive justice. It could be said, for instance, that it is not the role of a theory of distributive justice to allocate such goods as friends or sexual partners. Thus, talk of people suffering from undeserved inequalities in circumstances, and so on, needs to be interpreted in a context in which we are concerned only with specific kinds of social goods, such as income, economic holdings, political power, or political and civil rights.

Earlier in this chapter, however, I pointed out that it is difficult to draw a principled line between what falls inside and outside the class of benefits to be distributed in accordance with a theory of justice. Indeed, I suspect that there is considerable entanglement and continuity between the various kinds of competition that human beings enter into. But let us accept, for argument's sake, that a defensible line can be drawn. Even if we confine the discussion to competition for, say, money, jobs, and political offices, we usually accept the outcome when the person who has the greatest natural talent is, not coincidentally, successful. Of course, we look for something more than undeveloped talent. If the successful candidate actually has superior natural assets, though—in the sense of developed capacities and skills—all these can be traced back through upbringing and genes. We don't usually complain that the outcome was morally arbitrary because the person's capacities, skills, and so forth, were not deserved all the way down.

What Sorts of Advantages Don't We Accept?

Nonetheless, it might be said, we bridle at others' unearned or unfairly obtained advantages. This may well be true, but the paradigm of an unearned or unfairly obtained advantage is not just any advantage that can be traced back to natural talent, or something like early upbringing or good childhood nutrition. Speaking for myself, I don't bridle at all advantages that are obtained through abilities, resources, dispositions of character, and the like that are ultimately undeserved. I, and I suspect many others, resent more specific things. What things, exactly? I can sketch at least the beginning of an answer.

Consider all of the following. There are people who use their abilities and resources not merely to gain an advantage in some sphere where this is deemed appropriate, such as the labor market. Sometimes they use their capacities and skills, and perhaps their economic resources, to dominate others, acting with arrogance, pretension, or cruelty. There are other people who sometimes win out in competition, not because of some characteristic that we consider relevant and appropriate to the competition, but instead because of some extraneous characteristic or resource that we believe should not operate in this context. Examples include using bribery to win a court case or sports contest. Other people—or sometimes the same ones—seem to take what they want without appearing to make any effort or social contribution, perhaps simply by spending huge sums of money that mean nothing to them. They may strike us as being wealthy parasites.

If these are some of the things that many of us really object to, perhaps we have good justifications for doing so, even if we are not normally conscious of them. First, we have every reason to fear people with great abilities or resources if they use them in ways that are potentially detrimental to our interests. Second, we often have reasons to wish people to succeed in particular circumstances only if they confine themselves to a limited range of methods. We thus want a professorial post to be offered to somebody who has the potential to fill it with distinction, not to somebody who can obtain it by signing a check or somehow ingratiating themselves with the university president. We want sports contests to be won by team members or individuals who possess whatever capacities and skills we expect the competition to test—and not by someone with what appears an extraneous characteristic, such as wealth (which can be converted to bribes).

Third, and perhaps most important, we are frequently mistrustful and resentful when somebody is able to reach out with ease and obtain significant benefits with what seems like no effort at all. We want to feel that some advantages in life are *earned*—but not in the sense that they are earned all the way down (as if we could ever earn our own genetic potentials, uterine environment, early upbringing, and so forth). We merely want to see some kind of (significant) *effort* expended or some kind of (significant) *contribution* made to the welfare or projects of others. Certainly we are more accepting of individuals gaining large advantages if, in the process, they benefit society as a whole.

When we bridle at effortless gain, we may have good reasons to take a suspicious, potentially resentful attitude. If somebody is in a position to obtain significant goods without effort or contribution, this will have at least two undesirable effects: she will be able to gain in ways that involve no reciprocity with others, and offer no mutual advantage from interaction with her; and she may even be in a position to mock our own struggles and efforts, and perhaps acquire a position of dominance over us. This person is not a good one to share a society with.

A Better View of Desert

These points hint at an alternative conception of desert that could in principle be refined and elaborated at length. Its main features are that our concept of desert is, after all, compatible with the insight that human beings are not ultimate self-creators, and the concept is quite pluralistic. To pick up the first point, we do not usually object when somebody obtains a benefit by her own efforts and/or if she makes some sort of contribution to the welfare of others. We do not take a resentful attitude merely because her efforts may have been made possible (ultimately) by her natural talents or her upbringing. The sorts of benefits that we resent, and may condemn as "undeserved" or "unfair," seem to be more specific.

To elaborate the second point, we might ask, What is the "desert basis" (Feinberg 1970, 58–61) in a beauty pageant? What makes the winning contestant "deserve" to win? It might be different from what makes a football team deserve to win a game, what makes somebody deserve her wealth, or what makes somebody deserve a public award, criminal punishment, or civil compensation as the outcome of a lawsuit. In complex and subtle ways, desert is tied up with notions of effort, contribution, fittingness, and the exclusion of extraneous influences. In different contexts, many different sorts of things count for us as desert bases. As Kekes (1997, 124) puts it, desert "does not have a unitary basis; it is a pluralistic notion."

We can also see one reason why such an example as the beauty pageant appears troubling. Even leaving aside feminist concerns about women being valued or "deserving" of public recognition for their physical beauty (perhaps to the exclusion or undervaluing of more important characteristics), we sometimes feel uncomfortable when people of either sex use their sheer physical attractiveness to obtain an advantage. We might say that Miss Kazanistan deserved to win the pageant if we believe that she was the most beautiful contestant by some accepted standard. Certainly we'd say that she was not a deserving winner—in any sense of deserving—if the outcome resulted from some whim of the judges, some misunderstanding or national prejudice, or perhaps from nepotism or bribery. At another level, achieving financial and social gains through beauty sometimes seems problematic, or at least causes understandable resentment.

In part, this may be because beauty seems like something that takes little *effort* to obtain (for those who actually do obtain it) in contrast with the capacities, skills, and dispositions of character involved in sporting prowess, or successful (and honest) practice in a trade, business, profession, or academic scholarship. Beauty also can attract enormous social and economic rewards. When this happens, it may seem comparable to cases where people inherit huge fortunes; people who are blessed with sheer physical beauty, and who use it to obtain further benefits, sometimes seem to have too much handed to them on a silver platter. It seems too easy.

When do we deserve to have money? As an initial thought, we might say that somebody deserves the money they have if they are legally entitled to it and obtained it by honest, socially approved methods, even if it was actually a gift. Where a gift is involved, we may in fact hesitate somewhat to use the word desert and its cognates. But it does not follow that someone should not receive the gift, all things considered. We may feel that it is morally acceptable to give property to others as an act of love, kindness, or benevolence. Once the property has changed hands, we might say that the new owner is morally (as well as legally) *entitled* to keep it, even if they did not strictly *deserve* to receive it.

The possibilities can become complex and our reactions rather nuanced. Imagine that Diane now has a large sum of money gained merely from parental advancements or bequests, with no obvious effort involved on her part. In such a case, we *might* say, "Diane is entitled to (or has

every right to) the fortune her parents gave her," while *not* being inclined
to say "Diane deserved it." But what if it is pointed out that she put in
some special effort to be a "good" daughter, perhaps by a record of being dutiful and kind to her mother and father over decades, or perhaps
by caring for them in their years of need and frailty? Those patterns of
conduct may well seem like respectable desert bases after all.

If we adopt this more pluralist account of desert, we should conclude
that desert is not all or nothing: some benefits might be deserved to a
greater or lesser extent along a kind of continuum. Moreover, we may
often be ambivalent in our feelings about whether something seems to be
deserved or not. There may be factors pushing us both ways. Sometimes
we may be more comfortable using other language, such as that of merit
or entitlement (moral or legal), which may have different meanings from
the language of desert—though sometimes only subtly different.

At this stage, I can usefully emphasize a point that is crucial for any
discussion of human enhancement: a rich and plausible conception of
desert does not involve a claim that all undeserved benefits are injustices. We can conclude, for example, that many parental benefits given to
young children have not been *deserved* in any way, while also accepting
that they are socially expected and not, by our ordinary standards, unjust.
What we want is *not* that nobody ever be helped by their parents at an
early stage of their life. Rather, we expect (in a prescriptive sense of expect) that growing children will make the effort to develop capacities and
skills that originally existed only as potentials.

Assume that a naturally talented child, Erik, actually does make that
effort. As he develops capacities and skills, we then expect him to put in
further effort and make a social contribution. If in these circumstances
Erik's efforts gain him financial rewards, we are likely to consider the latter deserved, though the degree of comfort that we feel when we say so
may depend on just how much effort he has put in as well as the balance
between the social utility of his actions and his own personal gain.

Thus, we do not demand that our fellow citizens' developed or developing capacities and skills be deserved all the way down. It is true that
no child can deserve to have the genetic potential for an IQ of 150 or be
born into a caring, financially comfortable family. But that is not what
we demand of anybody. We are more likely to demand that children who
are fortunate enough to have such gifts as good genetic potentials, good
upbringing, good education, the advancement of parental funds, and so
forth, not squander them as they grow up and embark on adult life. More
specifically, we may demand that these fortunate individuals put in the
effort to convert their gifts (or at least some of them) into developed

capacities, skills, and virtuous dispositions of character. We *then* demand that they not use their gifts or developed capacities in ways that are antisocial or frightening, or that bring overly easy rewards, with no effort on their part; such individuals should (we may think) take on challenges commensurate with their powers.

Based on this view, the relevance of desert increases during a lifetime. There is little that newborn children could actually *deserve*, but there are good reasons to want them to be nurtured, educated, and socialized. By contrast, we might take quite a critical stance when we see grown and capable adults obtain windfall profits even if they do no positive harm. These are all, moreover, *moral* demands that are difficult to convert straightforwardly into *legal* ones. With all this nuance and complexity, we do not usually expect the law to allow only deserving people to obtain social goods. Unless some clear line is crossed, as when an important benefit is obtained by bribery, civil and criminal law are not useful instruments for enforcing such moral demands. This is partly because desert is usually a matter of degree; more than that, it may frequently, at least for practical purposes, lie in the eye of the beholder.

Finally, when we demand efforts and social contributions from our fellow citizens, it is not practical to specify approaches to life that are legally obligatory or legally forbidden. As long as they do not cause direct harm or anything relevantly similar, much of what we expect from other citizens is open-ended or vague. Often, as with what Immanuel Kant called duties of imperfect obligation, we must leave a great deal of discretion to individuals in choosing just how our moral demands of them should be met in their particular circumstances.

Desert and the Goods of Genetic Choice

Given all this, and depending on the situation, we might well take the attitude that Abigail deserves whatever money she uses to hire biomedical experts to modify the future Belinda's DNA. There is no more reason to suspect that the money was undeserved than would be the case with money spent on other goods and services available for Abigail's purchase.

But what about *Belinda*? Clearly the child did not deserve her good genetic fortune, but how much does it matter? Belinda has in fact obtained an enviable genetic potential through no prior merit or desert of her own, yet the same can be said of *any* child whose development is assisted by the good fortune of having a genetic potential to develop such characteristics as health, longevity, strength, and intelligence. No plausible theory of justice or desert can demand that genetic potential itself be deserved. Nor can any such theory plausibly claim that it is impermissible for parents to give their children undeserved benefits (out of love, care, or simply a wish that their children's lives go well).

Perhaps we would be justified in resenting Belinda if her genome were altered to such a degree that she would end up with extraordinary capacities as if by magic. We could imagine a scenario where some kind of genetic superscience made it inevitable that Belinda would attain godlike abilities without doing anything to achieve them. But nothing in the Abigail-Belinda scenario suggests that Belinda's developed abilities will end up being at anything like this level or that she will obtain them without effort. If she is to develop intellectual skills, she will need to study; if she is to develop high-level athletic capacities, she will need to make efforts in training, eating nourishing food, and probably making other efforts and sacrifices. Even a long life will not come to her unless she exercises, eats healthily, and generally shows prudence. While she will have greater *potential* than most, nothing beyond that will be handed to her on a silver platter.

Such considerations as these indicate that genetic modifications to boost favorable characteristics should be thought of much as we do such environmental interventions in a child's life as the provision of good nutrition and education. What we demand is *not* that children be prevented from having such advantages at all unless (impossibly) they already deserve them. Instead, we demand that they not waste these advantages. They should be accompanied by efforts aimed at developing capacities, skills, and useful dispositions of character. We also demand that as the child grows to adulthood, her developed capacities and skills should be used for social *as well as* personal benefit.

The Source of the Rawlsian Intuition?

At this stage I have criticized the Rawlsian intuition, briefly described what strikes me as a richer and more plausible account of desert along with its relationship to fairness and distributive justice, and suggested how this might be applied to the case of Abigail and Belinda. Still, some readers might have a lingering doubt. It might be thought that my criticisms of the Rawlsian intuition leave a mystery.

While there are good reasons to treat the Rawlsian intuition with suspicion, we might wonder why Rawls himself was so confident of it? Why, furthermore, is it shared by many other political philosophers? Writing on the closely related issue of economic equality, Frankfurt makes a pertinent observation. He acknowledges that some people do claim simply to have an intuition that economic inequality is morally wrong, but he expresses the suspicion that what is really disturbing them is a quite different point: that some are so *poor*. What we care about, he believes, is not equality but rather whether people have urgent needs to be met (Frankfurt 1989, 146–151).

Similarly, Raz suggests that we do not care about unequal distributions as such but other things instead: "the hunger of the hungry, the need of the needy, the suffering of the ill, and so on." That the hungry, needy, and ill have neighbors who are better off might encourage us to give priority to relieving the plights of those in hunger, need, and sickness, but we do so out of a sense of whose requirements are greater, not out of an abstract concern for equality (Raz 1986, 240). When we consider the plight of people whose lives are constrained by such things as hunger and need, we most likely wish to ameliorate their hunger and need for its own sake.

There may be more to it than this. For example, we also may be troubled that many differences in social status and wealth arise from past actions that we see as grossly unjust, such as acts of racial discrimination, economic exploitation, and so forth, and we may seek a political principle that acknowledges the current injustices in wealth distributions. This could lead us to the kind of reasoning that Buchanan and his colleagues refer to as the social structural view in support of radical measures for equality of opportunity. Again, what we care about is not equality as such but rather the effects of historical injustices. Finally, as already discussed, we may have various reasons for suspicion or hostility when we see windfall gains—when somebody is in a position to obtain significant goods seemingly without effort or social contribution of her own.

Such concerns about deprivation, historical injustice, and the merits of effort and social contribution may motivate us to look for a simple principle or "intuition" that appears to explain and justify how we feel. If the Rawlsian intuition were the only explanation of our feelings or how we could be justified in having them, there would be an argument in its favor. But that is not the situation. Our concerns about hunger, need, effort, social contribution, and the legacy of historical injustice do not guarantee that anything like the Rawlsian intuition will turn out to be an adequate starting point for public policy.

On the contrary, we have every reason to think that this simple intuition is not adequate to the complexities of our experience, our attitudes to the many situations we encounter in everyday life, or our real concerns that lead us to condemn some benefits (but not all) as undeserved.

Equality in a Liberal Society

The Social Structural View
To this point I have been critical of the Rawlsian intuition and the brute luck viewpoint that it might be thought to entail. In response, I've sketched what I take to be a more promising account of desert and its relationship to everyday intuitions. Yet there are other foundations for an egalitarian or at least redistributive element in public policy—foundations with a more concrete basis in human experience than the Rawlsian intuition. One approach that I have already mentioned relies on the perspective that current dispositions of wealth and power are tainted by past injustices of various kinds.

In developing such an approach, Buchanan and his colleagues begin with a look at equal opportunity and three conceptions of what it might require: the elimination of legal barriers to similar social and economic prospects for individuals with similar talents; in addition to this, the elimination of informal barriers, such as racial discrimination; and in addition to both of these, the elimination of social barriers, such as lesser wealth (this might require such initiatives as free basic education). They call the first two of these "nondiscrimination" conceptions of equal opportunity and the last a "level playing field" conception. The idea of the level playing field conception, which they support, is that being born into a family with low socioeconomic or educational status should not in itself give somebody lower prospects than other individuals with similar talents and abilities (Buchanan et al. 2000, 65–66).

Buchanan and his collaborators then explain two possible rationales for the level playing field conception: the brute luck view and social structural view (ibid., 66–73). The brute luck view is essentially the view of fairness that I have already examined and rejected. The social structural view is more promising. The idea here is that current economic distributions have been influenced not just by luck but also by the ongoing effects of unjust actions and practices, such as invasions, slavery, the dispossession of indigenous peoples, and less stark examples of unfair or exploitative actions by the rich, whose wealth is inevitably tainted to at least some extent. In a similar spirit, Howard Kahane suggests that those at the top of an economic pecking order generally receive far more, and those at the bottom far less, than what we would recognize as a proportional return for their efforts. Relying on proportionality between contribution and reward as his criterion for fairness, he thinks that few large fortunes are fairly earned (Kahane 1995, 79).

The strength of the social structural view is that it impugns only those advantages that can be traced back to actions that are universally condemned as unjust. Clearly this approach has some attraction: we often suspect and sometimes know that wealth has been obtained not merely with an element of luck but instead by methods that almost all of us deplore. As Buchanan and his colleagues hold, the effect of this has been pervasive, and one obvious response is to attempt—one way or another—to limit the degree to which it can affect subsequent outcomes. Thus, the social structural view is underpinned by facts about human experience. If we adopt the social structural view, moreover, that is sufficient for us to maintain that no one's opportunities in life should be advantaged by their initial *wealth* or the wealth of others such as their parents. Note that this relatively weak claim could be made without any commitment to the much less plausible claim that it is morally obligatory to prevent advantages that flow from *any and all* ultimately undeserved circumstances.

Nonetheless, there are some significant concerns if this view is relied on to prohibit certain uses of enhancement technologies. In that context, the social structural view runs into at least some of the same difficulties as brute luck accounts of distributive justice. The social structural view would not condemn some people's attainment of more success in life than others purely because of differences in natural talents. It could, nonetheless, condemn a very broad range of actions that involve parental assistance of one kind or another, whenever the assistance involves the use of wealth that the parents would not possess but for acts of clear injustice in the past. And it is arguable that *all* wealth falls into that category.

Yet many acts of parental assistance to children, including those that require a degree of parental wealth, are widely considered acceptable or even good. Furthermore, if the social structural view is relied on to suppress benefits rather than to level up opportunities to obtain them, it has the same problem as any other view that is used in an attempt to justify the elimination of benefits that instead should be extended more widely.

More fundamentally, the social structural view faces the problem that not all wealth creation employs unjust actions and practices, even though it is true that no actual holdings of wealth would exist in their particular form if history had taken a different course without the injustices of slavery, conquest, dispossession, exploitation, and so on. In fact, wealth creation frequently involves efforts that are widely thought of as morally legitimate and outcomes that obtain broad social acceptance. More than that, it can involve beneficial acts of creativity and leadership. No society could operate a system of property transfers if it assumed that all holdings of property were hopelessly morally tainted.

So how are we to sort out the mix of morally unjust and morally legitimate inputs into the creation of wealth? In some cases, perhaps many, we can clearly identify—even if we cannot quantify—the historical impact of (for example) slavery and colonial dispossession. In many ordinary cases involving people who have earned wealth through the exercise of effort and skill, however, it is not clear that we can ever disaggregate the effects of such past outrages, actions that are usually considered acceptable, and those that appear praiseworthy. All of these have interacted in the past, and there is no apparent way to calculate a determinate answer—no reliable way to unscramble the egg.

At the same time, there is a tension between the insistence on a strictly level playing field for social and economic prospects and the widespread desire of parents to obtain benefits for their children. Liberal societies generally accommodate this desire—undoubtedly, it is too powerful *not* to be accommodated by the moral norms of any society that allows the accumulation of property. Besides, some accommodation of this parental wish has beneficial effects, since it creates an incentive for citizens (at least those who are parents, or planning to be) to engage in productive work. Even if the desire of parents to obtain advantages for their children were universally deplored—which is far from being the case—it would be difficult to eradicate. Perhaps we are motivated, to some extent, to seek a "fair" marketplace in the sense of one where all participants in each generation start out on an equal footing. However, this conflicts with deep-seated desires that evolutionary psychology sees as originating in kin altruism—such as the desire to spend wealth on one's own children (Kahane 1995, 79).

If we do accept the private accumulation of economic resources, and if we also accept that wealth is often created and accumulated by legitimate or even praiseworthy efforts, then we have powerful reasons to permit at least some parental actions that involve spending economic resources to improve children's life prospects. In that case, we cannot rely on the social structural view to justify policies that totally remove the advantages obtained from parental wealth. Perhaps the social structural view can be used to justify policies that offer particular assistance to historically disadvantaged groups, where the lingering impact of past outrages is most clear, but it does not show that all well-meaning actions by the wealthy to assist their own children are unjust or that they ought to be prevented by the state.

Must we, then, accept that liberal societies should impose no restrictions at all on differential outcomes that result from parental efforts? No, that would be a premature conclusion.

The Human Experience

Neither the Rawlsian intuition nor the social structural view is sufficient to support the sort of fairness argument described earlier in this chapter. Nonetheless, liberal democracies may have many legitimate reasons for policies that limit the advantages conferred by family wealth. These rationales would apply even if all accumulations of private wealth were obtained entirely by morally acceptable methods.

For example, a liberal democracy might wish to ensure that as a society, it does in fact gain the benefits of its citizens' natural talents, at least to the degree that this can be done while also tolerating their deeply valued life plans. To illustrate this, the state may not ultimately prevent an individual from leading a life of religious contemplation or libertine abandon, or adopting the lifestyle of a nomadic surfer—in all cases, perhaps, making little economic contribution. Surely, however, it may take at least some steps in an attempt to ensure that its citizens' talents do not go unnoticed and undeveloped. This suggests a powerful reason for liberal democracies to educate all children of whatever socioeconomic background to a high level—or at least provide them with requisite educational opportunities. It tends to favor equal opportunity, but only by *upward* leveling. It is a reason that actually favors efforts to increase the initial talents of children by means of genetic engineering, if that proves to be possible.

Other reasons to keep inequality on some sort of leash include the decreasing marginal utility of greater wealth, the isolation and alienation of the poor, the difficulty of sustaining cooperation among people who are greatly unequal in their power to affect each other's lives, and the soul-destroying nature of a civilization that overemphasizes struggles for status (Singer 1999, 45–46, 52–53). Putting it another way, we may have broadly utilitarian reasons to redistribute wealth, reasons that relate specifically to the plight of the poor (and the hungry, the needy, and so on), and others that relate to social cooperation and stability (confirming that there is some artificiality in distinguishing considerations of social stability from those of distributive justice).

Consider the massive and destructive inequalities that are likely to arise in a political libertarian society that refuses to interfere with any pattern of holdings that eventually emerges from legally permissible transactions. People with greater natural assets will tend to obtain greater economic rewards than others for their activities and products, and the resulting economic inequalities will then be passed on to children. As wealth concentrates in relatively few hands, the effect is a hierarchy of significantly unequal socioeconomic classes. The effects of this, in turn, may be damaging for those at the bottom of society and for society as a whole. Potentially destructive feelings of envy will be aroused if citizens meet each other in ways that some find demeaning, or if some experience their situation as humiliating or impoverished (Rawls 1971, 536–537; 1999b, 470–471).

Hierarchical societies contain supposedly superior and inferior ranks, with an acceptance that members of the superior group are entitled to treat those of the inferior group with violence and other kinds of abuse. The inferior group may be segregated, enslaved, or forced to abandon its culture. Within such societies, individuals in the subordinated group are not able to stand as equals in public discussion, but are forced to bow and scrape, or represent themselves as inferior (Anderson 1999, 312–313). All of this is a recipe for suffering, discontent, alienation, and lawlessness.

Finally, consider the massively complex reality of the postindustrial West's large-scale, populous, economically sophisticated, technologically advanced, and largely anonymous societies. Without exception, they have discovered that relatively detailed social regulation is virtually unavoidable. It is required for the efficient operation of government and commerce along with many prosaic aspects of everyday life, such as access to public roads and highways. There is no serious alternative to active intervention by the state to redistribute property, provide a range of services, and generally coordinate interactions and ameliorate the plight of those who would otherwise be victims of a potentially cruel economic system.

Unfortunately, the considerations I've advanced do not all pull in the same direction. Some of them imply that we should limit the ability of parents to improve their children's life prospects, but others suggest allowing parents' efforts to have a beneficial impact. Concerns with facilitating individual autonomy and at least *tolerating* a wide range of parental ambitions for the futures of their children might favor allowing such efforts by parents. Furthermore, it should be acknowledged that such actions as educating children produce widely valued goods. It is understandable that parental efforts of this kind are commonly seen as praiseworthy, and that there is little prospect that this will ever change. There is a strong argument that no children should face significant economic barriers to developing their potential through education. But that consideration does not support any policy of *leveling down*. Rather, it is consistent with the broad policy approach adopted by modern-day liberal democracies—one of attempting to *level up.*

Meanwhile, egalitarian movements aimed at freedom and respect for identifiably disadvantaged groups have not pursued an agenda of exact equality of wealth or income for all citizens. Instead, they have opposed oppressive hierarchies in which, for example, men are seen as naturally dominant. It is not a matter of equalizing such things as wealth but rather of undoing relationships in which some are subordinated to others. This may in fact require extensive steps in the direction of equal opportunity for those with similar talents and ambitions, but strictly equal economic shares are not required. Nor is the relief of hunger, need, and poverty aimed at *equalizing* anything, though it may be important to maintain some limits on inequality. What matters is ensuring that wealth not be used as a weapon to create oppressive relationships of dominance and subordination, or to produce a class of people who are isolated, alienated, hungry, in need, or consumed by despair or anger.

Such an approach is better grounded in history as well as more appropriate for the goal settings of public policy in a liberal-democratic society than either a brute luck form of liberal egalitarianism or the more moderate social structural approach described by Buchanan and his colleagues. So why should a radically different approach to policy be adopted when enhancement technologies are under discussion? How are these particular benefits different from the many other useful things that money can buy, especially the many benefits that parents can buy for their children?

Revisiting the Arguments

The Fear of a Caste Society

Buchanan (2011b, 248–249) has highlighted the disagreement, difficulty, and current uncertainty that characterize debates over distributive justice, with no foreseeable prospect of a resolution. Throughout this chapter, I have put arguments that suggest that the more straightforwardly egalitarian theories—those that embrace the Rawlsian intuition—do not conform to our everyday thinking about justice. I believe that these theories are unlikely to prevail, whether philosophically or politically. In any event, Buchanan (ibid., 249) makes an attractive point when he notes that we must proceed in the face of uncertainty, adopting a relatively minimalist idea of justice on which there is much agreement. This leads him to concentrate pragmatically on reducing and avoiding extreme deprivations along with serious economic and political inequalities—so serious as to exclude people from a great of deal of political and socioeconomic participation (ibid., 249–252).

Whether for Buchanan's pragmatic reasons or those I've explored, straightforward arguments by political egalitarians cannot provide an acceptable basis for the prohibition of human genetic engineering. Far more needs to be said to demonstrate why this or related genetic technologies might pose a special, distinguishable threat to those who cannot afford them. The threat must not be merely that genetic choices can boost the relative prospects of some people (for so can many other things). Rather, it must be that those who miss out are thereby likely to suffer some unacceptable misery, deprivation, subordination, or exclusion, or something similar. What risks of this kind do we face, and what is the probability that they will eventuate if nothing is done? Why can't the risk be handled through appropriate policies to diffuse the technology or ameliorate its bad effects?

In the absence of a convincing case along these lines, it appears that a double standard is being applied when critics complain about the advantageous effects of genetic technology more than about those of "nongenetic innovations," as Holland (2003, 183) remarks.

One possible starting point is the fear that genetic interventions could bring such *great* advantages—to some people but not others. Those who are advantaged might receive benefits that are themselves valued, but in addition might confer unprecedented advantage in competition; early in this chapter, I suggested holding on to Glannon's claim that the advantages obtainable from genetic engineering potentially would have no limit. Furthermore, some forms of enhancement might be deeply transformational in their depth and permanence, and might make it easier to do more than one thing at a time, perhaps allowing some enhanced individuals to dominate across the entire range of social competition (Mehlman and Botkin 1998, 53–61). This kind of relative advantage would be difficult to reduce once it came into being, since genetic advantages, as opposed to the wealth they might produce, cannot simply be taxed. Great invasiveness would be required to reduce the disadvantage to those without access to the relevant genetic technology (Glover 2006, 79).

Unlimited and largely ineradicable benefits to *some* can produce harms for *others*—or so the argument might proceed. If these harms are sufficiently serious and cannot be avoided in other ways, then we should forbid whatever technologies are implicated or perhaps prevent their coming into existence.

Much of the existing literature on human enhancement deals with anxieties similar to these. Interestingly, Mehlman, one of the most prominent opponents of nontherapeutic genetic engineering, does not seek to forbid what he calls "passive" genetic enhancement, by which he means the use of PGD to choose potential traits. He points out that the characteristics of children chosen (as embryos) by PGD would remain within familiar limits, with no sudden, dramatic extensions of capacities. By contrast, Mehlman (2003, 171–175, 183–184) contends, active germ line modification of human embryos would have the potential to create a genobility and ultimately a race of gods.

Other commentators express concerns that the genetically superior class will come to see themselves as deserving of their mastery over the rest of us. Perhaps those who possess superior capacities as a result of choices by their parents, as opposed to chance, may feel more as if their talents are deserved and see themselves as a kind of justified aristocracy (Fukuyama 2002, 157; Sandel 2007, 89–92). In a worst-case scenario, genetically enhanced individuals might come to be regarded as morally and politically superior, with a claim to superior legal rights, leading to a caste society based on differential levels of genetic endowment (Resnik 1993, 32–36). If the capacities of some were enhanced to an extreme degree, they might treat the rest of us in ways that we'd find atrocious or oppressive, although morally justified by the oppressors' own lights (Agar 2010, 160–173). In any event, there could be a de facto, more or less permanent overclass (Kitcher 1996, 124–126; Silver 1999, 4–8, 281–293; Buchanan et al. 2000, 95; McKibben 2003, 38–41; Baylis and Robert 2004, 11–12).

One imaginative example of such a scenario is the extreme polarization of "Gene-enriched" (or "GenRich") and "Naturals" projected by Lee M. Silver (1999, 4–8) in his exploration of future possibilities emerging from PGD and "synthetic genes." Here the GenRich form an ultramodern hereditary aristocracy on the verge of becoming a separate species. Another such illustration is the 1997 science-fiction movie *Gattaca*, directed by Andrew Niccol. In contrast to Mehlman's relatively accepting (or resigned) attitude to PGD, *Gattaca* depicts a caste society that has come about by means of widespread PGD to select desired traits. When this is combined with the relentless use of a genetic database and associated testing to guide social decisions, the outcome is a subordination of the interests of "in-valids"—those conceived and born without the use of PGD.

These are nightmarish scenarios, and averting them does seem like a legitimate exercise of state power, but are they realistic possibilities?

Assessing the Nightmares

The question that confronts us, then, is not whether the benefits of enhancement technologies can be made available at the same level and same point in time to everybody. Let us grant that this is not a realistic prospect. Rather, we must ask about the realism of speculations that the availability of human enhancement will lead us all down a path that culminates in a morally intolerable caste structure or something that approaches it too closely for comfort.

Unfortunately, the question must be addressed without the benefit of empirical data. We cannot know what technological applications will actually become available in the absence of attempts at suppression, or how effective they will be in achieving their purposes. Moreover, even if they prove to be expensive, we don't know whether they are likely to be adopted by a large stratum of wealthy citizens or only a small group— perhaps just a scattering of rich eccentrics too few in number to oppress the remainder. That said, I do not believe we're entirely helpless when we reflect on the possibilities.

Some advocates of enhancement technologies see the prospect of a caste society as totally unrealistic. Among these, Ray Kurzweil argues that current economic inequalities will be irrelevant to any social outcomes from new technologies because we will attain a world of plenty. He refers to information technologies, which (he insists) typically begin with inefficient versions "that are unaffordable except by the elite"; the technology then improves, while becoming "merely expensive"; later it "works quite well and becomes inexpensive," and eventually "works extremely well and is almost free" (Kurzweil 2005, 469).

Kurzweil may be correct to this extent: such a pattern may occur frequently, applying to many new technologies under many social conditions. I am unwilling to assume, though, that it will apply to all the technologies that concern us here. Kurzweil's main example is cell phones, but what about cosmetic surgery? While this is more widely available than was once the case, it cannot be assumed that it is at the "knee" of a curve of exponential change, and eventually will be highly effective and available to all.

No deeper reason has been given as to why technological diffusion must always follow the same pattern, and we have limited experience to date with powerful, intrusive technologies that can alter the human body and its capacities. It may be that the diffusion of new technologies typically involves a period with a sharply accelerating rate of adoption, with the overall rate over time eventually flattening out into an S-shaped curve. But many factors can affect the rate of adoption for different technologies as well as the percentage of the population using them when the curve finally flattens out. These can include the expense of the technology (whether measured in money, effort, risks, or other burdens imposed on users), the wealth or otherwise of users in various markets, and whether the technology's popularity is increased by beneficial network effects. Accordingly, any attempt to predict the uptake of enhancement technologies is likely to be quite inconclusive, as with a recent discussion of the subject by Anders Sandberg and Julian Savulescu (2011, 99–102).

We should not be content to rest a defense of human enhancement on Kurzweil's model of technological development and diffusion. This seems too complacent. On the other hand, we cannot simply assume that whatever human enhancement products become available in the capitalist marketplace will be extremely powerful and (for most people) prohibitively expensive.

It is easy to imagine a scenario, for example, where enhancement technologies in the sense discussed in this study (reproductive cloning, PGD, and genetic engineering) do not actually become widely available as practical options, but related products do—perhaps pills of certain kinds that can boost some capacities. These might be the result of research related to human enhancement, and might be far cheaper and more common than custom-designed genetically engineered embryos. Even the prenatal enhancement of embryos might eventually be conducted by some cheap and noninvasive method akin to taking a pill (Fox 2007, 6). It is also possible to imagine that such methods will have only limited efficacy, with the experience and effects resembling more those of good nutrition than those of the transformational events depicted in superhero comics (discovering a magical crystal, say, or being bitten by a radioactive spider).

Again, consider the expense involved in IVF. This might be seen as a barrier to the routine use of enhancement technologies and their widespread availability through all social classes. Yet it is telling that some critics of human enhancement fear that it will become available only to the rich, because of the expense, while others deplore the possibility of a future society in which nobody reproduces in the traditional way and IVF is the universal means of procreation. In chapter 5, I questioned whether the latter outcome should really cause us concern, however strange a society like that might seem to people who did not grow up in it.

As for the fear that the expense of IVF will pose a barrier, and thus divide society into superior and inferior ranks, this may prove to be a mirage. Social change has allowed—and to a significant extent pressured—women to have their children, if any, at much later ages than was once the case. This, in turn, creates a pressure for new technological products. In vitro egg maturation could supply one response. If young women could provide a one-off sample of immature eggs for freezing, followed by IVF at a later date, this could eliminate much of the discomfort and expense from IVF (Green 2007, 51–52). It would cut out the burdens of superovulation, perhaps greatly reduce costs, and make IVF more accessible and user friendly. It might become almost universal.

The result would indeed be a greater separation of sex and reproduction, but that is not a good argument against it. In a future society in which the use of in vitro egg maturation plus IVF has become the norm, there might be no serious class barriers to the use of PGD or even genetic engineering.

Such speculation aside, there are some things of which we can be reasonably confident. Enhancements requiring the insertion of nonhuman DNA sequences will usually be difficult to achieve, even by the most unscrupulous Frankensteinian scientist, and any that are easier to achieve (skin that glows in the dark, perhaps?) might not be useful in social competition. In the tightly regulated world of biomedical research, the required experiments may be almost ruled out for safety reasons. Furthermore, it will be far easier to improve average or below-average performance than to produce superhumans; for instance, it would be difficult to produce an adult nine feet tall with no grave health problems, as even McKibben (2003, 14) concedes.

Thus Stock is surely close to the mark when he argues that "strange fringe interventions," perhaps involving wings or gills, are not likely. It is plausible, as he suggests, that it would take something "extremely desirable"—such as extended longevity—as well as safe and reliable to obtain commercial viability. Stock (2002, 60–64, 180–181) foresees the likelihood that we will raise health and average capacities, probably compressing rather than flaring out the existing differences.

Even if Stock is too relaxed about the probable outcomes from the widespread use of enhancement technologies, the benefits without limit that Glannon writes of are not likely to emerge. The actual enhancements that are likely to become available will be limited by many factors, including safety, practicality, the undoubted dangers associated with introducing transgenic genes into our children, and the limited horizon of human desires. Parents may prefer to have children much like themselves, but with some enhanced capacities, rather than to become the progenitors of gods or monsters with which they will have little in common.

There are other reasons why the fears of a genobility may not be realistic. Consider the worry that those with technologically boosted capacities will come to see themselves as deserving in a way that is not true of people whose superior cognitive or physical powers result more directly from chance. Surely this fear is hard to credit; while none of us are self-creators all the way down, a genetically engineered child such as Belinda will certainly not be *more* of a self-creator than an ordinary child. If anything, one might expect the psychological impact to go in the opposite direction: if our natural talents (or those of a privileged group) became less natural, in the sense of more controllable by human choice, it might at least make more *obvious* the fact that no such assets are deserved in an ultimate way. Additionally, if some taint attaches to family wealth and privilege (as suggested in the social structural model of distributive justice), it would likewise attach to capacities and skills paid for, in part, out of family wealth.

Green contrasts the current situation in which fortunate people ignore the ultimate contingency of their good fortune and regard themselves as self-made, taking credit for their own hard work (and possibly even considering their superior abilities to be blessings from a providential deity). If superior abilities were more obviously a result of choices by other human beings—such as parents—this might tend to soften attitudes of self-righteousness about abilities and achievements, and encourage the sentiment that genes for health and fitness should be made accessible to all (Green 2007, 156–158). To whatever degree people can be induced to think like Rawlsians, or at least like social structuralists in the manner of Buchanan and his colleagues, the tendency will be to subvert the perceived legitimacy of unequal competitive outcomes and increase rather than reduce feelings of solidarity with the less fortunate.[3]

All that said, a residue of concern remains. The danger, as I see it, is not that genetic gods or some gene nobility will emerge. No prima facie case has been made out that anything so dire would happen or is even likely. But there is still a risk that the widespread use of genetic engineering and PGD will exacerbate existing class divisions, at least to some extent. We don't need to entertain the most bizarre and nightmarish scenarios to appreciate the likelihood that these technologies will be more available in practice to wealthier middle-class families than to those in less privileged social strata.

The barriers to accessibility may include elements of expense and inconvenience that are sufficient to deter many parents. Even if some of these are overcome—for example, if some genetic choices become cheaply available—there still could be a skew in their use toward wealthier groups. Over time, we could see an outcome where high levels of longevity, health, intelligence, and natural talents become more prevalent within higher socioeconomic groups ... but not so much so in lower groups. Simultaneously, low levels of these goods might be largely eliminated among the more privileged socioeconomic groups. The result need not be anything like a genobility, but it might add an extra degree of rigidity and visibility to social stratification.

In the extreme, an overall increase in intelligence levels that is strongly skewed toward higher socioeconomic strata could be harmful to those people who are not benefited. Such people might be locked out of many of the productive activities of society (cf. Wikler 2009). Even if nothing so dramatic happened, much less anything like the more outré speculations that are rife in the antienhancement literature, we could have somewhat more hierarchical societies. Though this is rather speculative, it seems to be a realistic picture of human enhancement's possible downside.

The Role of Government

Enhancement technologies could, among their other effects, push contemporary societies somewhat further in the direction of socioeconomic hierarchy. It is of course not unique in that respect: *many* technologies and practices can have that kind of antiegalitarian impact to a greater or lesser extent, and this is not usually seen as a good reason to enact prohibitions. How, then, should governments respond?

Given the benefits that enhancement technologies can bring to citizens and entire societies, governments have good reasons to develop sophisticated policies that are reasonably enhancement friendly. Most obviously perhaps, the governments of the future can ensure that the possession of enhanced talents and other beneficial traits does not confer greater political or legal rights and liberties, even if it inevitably bestows some advantages in social, economic, and sexual competition (Resnik 1993, 36–37).

That, however, is a river yet to be reached, much less crossed. At this earlier stage, governments can push back against the antiegalitarian impacts that so often accompany social and technological innovation by adopting reforms to reduce economic inequality. The aim here is not to produce perfect equality of resources, or destroy all advantages from upbringing and natural talents. It is something more modest: to employ taxation and spending programs to counteract the many pressures that tend to exacerbate social inequalities.

A government or citizenry that fails to commit itself to that approach more generally may be able to act opportunistically to suppress or hinder some specific innovations if they are unpopular for other reasons. That approach, though, is unlikely to achieve real success in counteracting pressures toward inequality. At the same time, it tends to corrupt public policy and undermine the idea of liberal tolerance whenever laws are enacted on illiberal grounds. That remains so even if the prohibitions also serve an ulterior purpose of counteracting inequality.

Should powerful technologies start to emerge, specific steps could be taken to spread the benefits beyond a particular social stratum—much as happens with other goods such as literacy. Once again, deliberations to produce the best policy mix would be affected by such imponderables as the timing of various innovations and perceived benefits. As I discussed briefly in chapter 6, there might be at least a limited role for the kind of state-funded lotteries proposed by Mehlman and Botkin. An element of chance could also be introduced into the eligibility to participate in trials of new products. But I expect that governments will be able to develop policies far more sophisticated than this as long as the technology is introduced relatively gradually, allowing time for political decision making.

I must also emphasize that issues relating to antiegalitarian pressures are, at the end of the day, of relatively little relevance to many controversial uses of enhancement technologies. Some applications might become differentially available to the rich, yet have limited tendency to assist them in outcompeting the rest of us. Examples include reproductive cloning, sex selection, and the selection of relatively trivial characteristics like eye color. Whatever else might be said against such innovations, they are unlikely, in themselves, to offer anyone much assistance in socioeconomic competition.

Something much more is required for that; at a minimum, powerful technologies that seriously skew the distribution of natural talents are necessary. This could apply to some futuristic uses of PGD and genetic engineering based on highly sophisticated understandings of the human genome. The scenario depicted in *Gattaca* raises the possibility that the comprehensive use of PGD merely for health and disease propensities would be enough—but how realistic is this?

Gattaca depicts a society in which the so-called in-valids are deprived of economic opportunities based not on evidence of their developed abilities or virtues of character but merely on their supposedly inferior genetic potential. They are victims of a kind of irrational discrimination that is not in the interests of potential employers or their society as a whole, which is denied the use of their developed abilities, even when these might be superior to those of their genetically selected rivals. This is an extreme version of what *could* happen even now with genetic testing: in principle, a wide range of genetic tests for disease potential could be used to discriminate against a worker who nonetheless might have a superior ability to perform the duties of the job.

As genetic testing grows more powerful in its scope and more precise in its predictions, it may play a greater role in employment decisions. There is a possibility of the ignorant misuse or outright abuse of genetic tests, and this will require a regulatory response to protect private information and counteract unreasonable forms of genetic discrimination in personnel management decisions. In a future world with differential access to PGD, similar regulation would no doubt be needed to prevent discrimination against "normal" people, who might be assumed, thoughtlessly and unreasonably, to have lesser potential as employees. It might also be needed to protect people who are the products of enhancement technologies, and therefore might be thought of as somehow being "cheats" or just "creepy."

Nevertheless, as Colin Farrelly notes, vague fears of creating a genetic underclass by such means need to be given more development than the critics have managed to date. This includes the need for an answer as to why a mix of some availability of enhancement combined with the prohibition of genetic discrimination would not help avert the formation of a genetic underclass (Farrelly 2004a, 591–592). That is particularly the case if we go down the path suggested by Buchanan (2011b, 243–278), who sees the problem as part of a wider one to do with the diffusion of beneficial technologies between as well as within societies. More work is needed to design mechanisms to achieve this.

Conclusion

One concern raised by the emergence of enhancement technologies is that new kinds of biologically based inequality will ensue, producing a radically unequal society in which some individuals suffer demeaning circumstances or other harsh outcomes. The concern need not be based on any controversial theory of distributive justice. It is sufficiently worrying that we might see the emergence of a hierarchical social structure, leading to morally intolerable relationships of dominance, subordination, and exclusion. Anything like this will involve harms of a kind closely analogous to others that modern liberal democracies try to prevent—notably the cruel outcomes for some from the workings of capitalist competition.

Once the discussion is framed in such a way, however, consideration needs to be given to the nature, likelihood, and probable extent of the risk, what can be done to avert it given our limited knowledge of the future, and the range of possible regulatory responses. We should try to avoid a situation where averting a possible set of harms also prevents many benefits. Where at all practical, we should identify responses that favor individual liberty and autonomy, allowing a wide range of life choices.

8

Policy Implications

Throughout the previous chapters, I have examined a range of arguments against enhancement technologies. My focus has been on contentions with plausible liberal credentials, referring to such concepts as harm, autonomy, and justice. In chapter 5, I explored the specific possibility that change could take place with such unprecedented speed that many people could not reasonably be expected to adapt to it and would find it harmful. Such rapid change has certainly not, however, been the experience to date. As I write, sixteen years after the cloning of Dolly was announced, there have been no credible reports of human reproductive cloning. The genetic engineering of human embryos to enhance beneficial characteristics seems faraway indeed.

The arguments overreach, but some contain a kernel of truth. There is, for example, a theoretical possibility that social stability might be threatened in one way or another by the prospect of human enhancement through genetic engineering. Again, my analysis to this point leaves a lingering fear that at least some genetically engineered people of the future might suffer psychologically from the knowledge that their genomes had been deliberately modified. And again, the rate of technological and social change *might* yet become intolerably rapid. But these considerations are largely speculative, and provide no good grounds for widespread departure from liberal tolerance.

On the contrary, a cool examination of the issues indicates that there might be much benefit—to parents, children, and society as a whole—if a range of enhancement technologies and applications became available. They could offer new reproductive options for would-be parents, and increased physical, perceptual, and cognitive capacities for children (perhaps including an increased capacity for self-creation and reflective self-scrutiny). A future society might benefit from the abilities of children whose genetic potentials were chosen by PGD or boosted by genetic engineering. The benefits enjoyed by its individual citizens might, in at least some cases, be greater the more widely shared they became; there is no reason to look on augmented human capacities as mainly positional goods. And far from causing tangible harm, in Devlin's sense of this expression, enhancement technologies might broaden and deepen a society's pool of available talent, and lead to benign network effects.

In this final chapter, I will address various residual concerns. In response to them, I'll propose principles that should be applied to the development of regulatory policy within a framework of liberal tolerance and legislative restraint.

Residual Concerns

Glover has made one attempt to work out the desirable limits to enhancement technologies. His suggested policy goals include preventing social inequalities from becoming more deeply rooted genetic ones; avoiding competitive traps based on merely positional goods; protecting the rights of children to an open future; and containing our darker nature or protecting our nature's good side (Glover 2006, 103). These notions do indeed reflect some, though not all, of the concerns that I've identified throughout this book. For example, if we could dismiss some benefits as merely positional ones, the cost of achieving them would give us genuine cause for concern. Yet as discussed in chapter 3, it is not so easy to identify benefits that are *merely* or *solely* positional, and some important enhancements of capacities might produce more benefit for individuals the more widely they are possessed by others.

Again, there is merit in the wish to protect the openness of children's futures. It appears that some genetic interventions could actually offer *more* open futures, however, rather than less.

It is worth pausing to review this crucial point. Some would-be parents might attempt to produce a child with a minor or not-so-minor disability, making the child unusually dependent on assistance from others and hindering her ability to take part in (some) activities available to most people. Again, a parent might seek an unusually docile, deferential, or credulous child—perhaps all the better to instill the parent's preferred worldview or system of culturally sanctioned values, or make the child submissive and dependent. But it need not work like that. It is instead likely that many parents would wish to confer the genetic potential for such characteristics as superior intelligence, memory, musical talent, beauty, strength and stamina, longevity, resistance to disease, or perceptual acuity. If actually brought to fruition during the child's development, many of these potentials could open up life plans that would otherwise be closed.

In many cases, the realistic threat to a genetically engineered child's future might come from outside rather than from the biological interventions themselves. It might, that is, come from the prejudices of others. Alternatively, it might come from the child's own reaction, possibly misguided, to the knowledge of having a human "designer." As to the latter, none us are self-creators all the way down. It is nonetheless conceivable that the knowledge of having a deliberate design, produced by other human beings, might be psychologically disturbing. It might be more disconcerting, at least for some children, even than such metaphysical and scientific theories as hard determinism (or "merely" compatibilist free will), or the denial of human exceptionalism. Much of this is highly speculative, and there is a danger of engaging in armchair psychology. Still, the most extreme and troubling cases would seem to be those where the would-be parents have engaged in thoroughgoing genetic engineering for exactly the "product" they want, or they attempt to control the child's personality and values directly. These cases can be contrasted with others where somebody's boosted physical or cognitive capacities have indirect as well as unpredictable effects on the development of their personality.

As alluded to above, Glover identifies the additional fear that existing social inequalities could become genetic ones. We can even imagine scenarios in which a hereditary caste of genetically superior individuals lords it over a despised underclass of genetic "naturals." The latter might suffer from diminished welfare, restricted life chances, political marginalization, and perhaps conditions of shame, squalor, or slavery. They might live with no hope of anything better for themselves, no matter how much talent they may have (at least by *our* standards), and no matter how much effort they apply. These are legitimate fears, although I argued in chapter 7 that they are unlikely to come true in an extreme form.

Finally, Glover writes of containing the dark side and preserving the good side of human nature. I examined similar concerns in chapter 6, where the emphasis was on outcomes that could be seen as harmful to society overall as opposed to just the individuals most immediately affected. The ability of human beings to live side by side, with all the benefits that ordered societies typically offer to their citizens, is highly dependent on our psychological makeup. We have evolved as social animals, and our success in building and maintaining harmonious societies relies on our built-in capacity for affective communication, responsiveness to others' distress, and preparedness to compromise our individual interests for collective benefit. Even our flourishing as individuals may depend on a built-in capacity to develop certain virtuous dispositions of character, such as kindness, tolerance, and a capacity to love other people for who they are. Our success as a social species and as individuals is based on this psychological capital. If we squandered it by making unwise changes to our children's natures, the outcome could be tragic.

In chapter 7, I noted that there is an artificiality about separating issues of social stability from those of equality. Neither concern simply outflanks or subsumes the other: social stability could be weakened by developments other than inequality; conversely, some societies might remain stable for long periods while displaying inequality in extreme and harmful forms. All the same, there is a connection: some forms of social inequality can provoke widespread resentment and threaten to snap social bonds (Rawls 1971, 536–537; 1999b, 470–471). Hence, the possible threat to social stability provides *one* basis for concerns about serious inequalities.

A Reality Check

When reviewing all these concerns, we ought not lose sight of the more immediate problems associated with the development of enhancement technologies. These problems forcibly underline the technical difficulties and also cast doubt on why it has been considered necessary to enact extensive legislative prohibitions in so many jurisdictions.

The first and most immediate problem is that experiments with reproductive cloning or genetic engineering are likely to create seriously malformed children, if children are actually brought to birth. This has enabled some opponents of cloning, such as Kass (2001, 34), to suggest that medical researchers cannot ethically make reproductive cloning safe or even find out whether it *can* be made safe. The idea is that any program of sufficiently decisive experiments would inevitably breach accepted ethical standards. For the experiments to be decisive, it would appear necessary to bring to birth children created by SCNT, yet it would not be ethical to bring such children to birth until the techniques had already been tested by decisive experiments. Catch-22!

There are further problems if we propose to use genetic engineering for any purpose beyond the therapeutic correction of specific disease-related DNA sequences. If we hope to "edit in" desired traits or enhancements of beneficial traits rather than merely "editing out" the potential for disease, the situation is discouraging. In most cases, predicting the outcome of interventions aimed at editing in is highly problematic, because of the pleiotropic character of many genes and the polygenic nature of many phenotypic traits (e.g., President's Council on Bioethics 2003, 39–43). The effect of this is that a technology of safe genetic engineering, at least for purposes that go beyond relatively simple kinds of genetic therapy, may lie beyond our grasp indefinitely.

Note that even if genetic engineers confined themselves to producing carefully chosen combinations of the three billion base pairs, including identified variations, that constitute the human genome, such problems would arise. Controlling and perhaps enhancing any genetic potentials of great interest or significance would be an extraordinarily difficult task. Perhaps there are combinations of single-nucleotide polymorphisms (SNPs) that *could* occur naturally and would produce highly desirable combinations of potentials—perhaps for intelligence or longevity—without detrimental effects. Identifying these would be difficult, however, without first conducting experiments to see the outcome. Even if armed with complex computer matching of observed traits in large populations, correlated with combinations of SNPs, genetic engineers might be poorly positioned to produce truly novel and dramatically beneficial modifications to an embryo's existing genome.

But it does not end there. The task is more complex again, and even less predictable and safe, if we hope to use more drastic techniques in an effort to produce spectacular changes to an embryo's potential. We might, for example, try introducing changes at points in the embryo's genome that are not usually variable within the human population. We might introduce nucleotides that do not frequently occur in nature at those points where variation is usually found (i.e., at the small percentage of base pairs where SNPs happen). More dramatically still, we could engage in transgenic engineering by inserting strands of DNA from nonhuman animals. These radical techniques could in principle create human or humanlike beings with a vastly extended range of morphologies, capacities, and other traits. In some cases, their abilities might lie beyond the historical human range. In particular cases, though, the unpredictable interactions of genes might well produce something unexpected and undesired.

It is worth emphasizing these practical hurdles to reproductive cloning, as well as to any ambitious attempts at genetic engineering, in order to maintain the realism of the policy discussion. Some technical or conceptual breakthrough would be required before genetic engineering could produce dramatic, beneficial alterations to the human genome. Without this, and independently of any regulatory action that might be taken by the state or professional organizations, researchers will be deterred from experiments that involve bringing genetically engineered human children to birth. I expect that *some* dramatic transgenic changes will turn out to be scientifically possible. We have seen this with certain experiments on monkeys, where a gene encoding for a fluorescent protein has been successfully edited in (Sasaki et al., 2009). But there would be great difficulties in producing transgenic variations to the human genome that are truly useful, with no downside.

For the foreseeable future, the genetic engineering of human embryos or babies is likely to be rare. The most obvious exceptions will be babies whose DNA has been modified to restore a single gene, or perhaps a small combination of genes, from a diseased state to what is considered normal. Moreover, there will often be an obvious and simpler alternative to genetic engineering. In many cases where parents realize that they are at risk of producing a child with the genetic potential for serious disease, it will be more practical to employ PGD to select an embryo without the disease potential. This type of selection might even become the rule rather than the exception in a future society in which sex and reproduction were typically kept separate. By itself, however, it would not cause any dramatic alterations in human capacities.

When the issues of safety and efficacy are taken seriously, some of the other concerns begin to fade away, because the level of technology required to make them realistic prospects appears to be out of reach. This highlights the fact that legislative bans on reproductive cloning and genetic engineering have to date actually had little work to do. Any experiments to bring to birth human children created by SCNT and/or modified by the actions of genetic engineers would already have been ill advised— and proscribed by ordinary ethical standards applicable to reproductive medicine.

In reply, it might be suggested that such standards lack teeth; if they take the form of mere guidelines, breaching them might not have sufficient consequences to provide an adequate deterrent. This may or may not be true depending on the formal status of ethical guidelines in particular jurisdictions and what consequences are attached. Ethical guidelines can, of course, do more than offer advice. Governments can attach funding penalties (applicable to government-funded researchers and their institutions). Professional bodies can be required by law to take disciplinary action in the event of serious breach. Courts can be directed to refer to the content of ethical guidelines when deciding civil litigation (such guidelines provide one standard for reasonable medical practice). Legislatures can even enact laws stipulating that certain breaches of so-called guidelines are civil or criminal offenses.

In any instance where there is genuine fear within a particular jurisdiction that existing standards for medical practice and biomedical research are inadequate to deter maverick human cloners or genetic engineers, there is an obvious response. A government-sponsored review could be conducted to consider the content and legal status of whatever ethical guidelines or similar regulatory instruments are used in the jurisdiction. Any such review could include the question of whether emerging theoretical prospects, such as reproductive cloning, bring new temptations to conduct unethical experiments. If required, regulatory provisions could be tightened up, given greater legal force and legal consequences, and/or extended in their coverage of sectors and industries. Given this alternative, there was never a need for draconian legislation creating entirely new crimes.

Problems of safety and efficacy admittedly would not deter *all* practices that provide for genetic choice. In particular, for all I've said, PGD can already be used for various purposes, including sex selection. Moreover, a wide range of research can proceed with no prospect of bringing to birth a child whose phenotypic characteristics would be radically unpredictable (bear in mind that a radically unpredictable phenotype is likely in practice to be one that involves debilitating deformities or is not viable at all). But that raises an important point: Why not allow less obviously risky forms of research to go ahead without hindrance?

PGD and Detrimental Traits

One legitimate concern about PGD is its possible use to select detrimental traits. In chapter 2, I discussed the difficult policy issues that arise when we contemplate the prospect that a child who would not otherwise exist might be brought into existence with, say, profound deafness. Where the state exercises political coercion to prevent this, there is an argument that its action goes beyond the harm principle, since (strictly speaking) no one is *harmed*.

Yet I argued earlier in this book that liberal democracies might legitimately take an interest in ensuring that PGD is not used to select embryos with the genetic potential for an identifiable and more than trivial disability. In essence, I maintained that all societies have good reason to favor the flourishing and success of their children. It does not follow that, all things considered, the state should enact prohibitions. For example, it might be best, again all things considered, to tolerate a few well-meaning actions to bring deaf children who would not otherwise exist into the world. While deafness is a significant disability, its effects are considerably mitigated in at least some circumstances—so much so that it is genuinely controversial whether, in those situations, it truly impedes a child's life prospects.

But what if it turned out that some parents were using PGD in an attempt to create *blind* children, deaf *and* blind children, or children with cystic fibrosis or Huntington's disease? If my arguments in chapter 2 are cogent, the state would be justified in using coercive power to deter this. Although this is, strictly speaking, an extension of the harm principle— not a mere application of it—the extension is consistent with liberal ideals and not merely ad hoc.

A Missed Opportunity

We live in a time of changes, and we will doubtless be faced in the near and more distant future with fellow citizens who adhere to new or unpopular belief systems, wish to lead unusual ways of life, plan to adopt technologies that meet widespread disapproval, or are motivated to express themselves in ways that others find challenging, shocking, or offensive. When confronted by all this, the starting point for public policy in a liberal democracy should be the value of tolerance.

We may have good reasons to consider the cumulative effect of permitting some actions—with potential outcomes such as social instability or the emergence of a hierarchical society with harsh effects for the "losers." On the other hand, we should not assume that the cumulative outcomes will always be negative. We most certainly should not underestimate the capacity of modern societies to accommodate a wide range of activities, systems of values or beliefs, ways of life, forms of self-expression, and styles of presentation. In the absence of direct, significant, wrongful harms to secular interests, we do well to apply skeptical scrutiny whenever we are confronted by arguments for directly coercive political action.

In any area of public policy, the idea of liberal tolerance is one that we should abandon, confine, or water down only with great reluctance. Unfortunately, the cloning of Dolly back in 1996 was a missed opportunity for the world's liberal democracies to reaffirm this. The opportunity that presented itself, and that was missed, was diametrically opposite of what Kass has claimed. According to Kass (2002, 173), the danger of human reproductive cloning provided an opportunity to "strike a blow for the human control of the technological project, for wisdom, for prudence, for human dignity." This blow would be struck by the enactment of legal prohibitions. I see things entirely differently.

A genuinely liberal approach to the prospect (remote, as it turned out) of human reproductive cloning would have been focused on issues of safety and efficacy, and on the ordinary ethical issues that arise when vulnerable people find themselves looking for medical solutions to their problems. There was no need for draconian laws calculated to demonize and repudiate the whole idea of reproductive cloning. Instead, there was an opportunity to make a powerful symbolic gesture and create a compelling precedent for the future. In the absence of good arguments grounded in the harm principle (or well-reasoned extensions to it), Western liberal democracies could and should have refused to ban reproductive cloning.

Instead of a calm and tolerant approach to public policy on enhancement technologies, we have seen highly intolerant reactions. Policy debate on these issues has been distorted by fears of violating the natural order, playing God, or undermining our sense of life's meaning, and by highly contestable conceptions of autonomy. It is, I submit, an understatement to observe that these are not considerations that ought to drive regulatory policy in any society with liberal pretensions. Where public opinion or powerful lobby groups favor the use of political coercion, that is all the more reason for a measured response by legislators, including public justification of the merits of liberal tolerance in the face of change.

Unfortunately, there is often a pressure on legislatures to be overly cautious about technological innovations. Legislators can be tempted to adopt negative attitudes when confronted by actual or potential innovations that frighten lobby groups or the public, perhaps raising fears of playing God, violating the natural order, and so on. The temptation is to set aside the merits of liberal tolerance and appear to be *doing* something.

Legislatures are faced with strong, well-organized lobbies that do not attach prime importance to liberal tolerance but rather their own religious, metaphysical, or moral agendas. Writing in a slightly different context, Green recounts his experience in the United States during the mid-1990s as a member of the Human Embryo Research Panel of the National Institutes of Health. He describes the powerful influence of religious lobbies in persuading Congress to reject any recommendations to allow federal funding of research related to human embryos (Green 2007, 205–206). In this case, the religious lobbies achieved a continued ban on federal funding, not outright criminalization of research, but the story confirms the political influence of well-resourced lobby groups with little commitment to liberal tolerance or the harm principle.

At least for the foreseeable future, it may be too late to undo the damage to an ethos of liberal tolerance caused by outraged reactions to the merely theoretical prospect of human reproductive cloning. An opportunity to exercise liberal tolerance and the political restraint that it entails—and reaffirm key liberal values—was squandered during the 1990s and the early years of the new century. Worse, a publicly conspicuous precedent was created for enacting directly coercive laws whose primary motivations are essentially illiberal or antiliberal. As a result, it will be even more difficult to argue for tolerance and exercise legislative restraint the next time some technological or social innovation excites widespread fear and repugnance.

Is there a way to avoid such outcomes? In the following sections, I propose principles that might guide policymaking for elected officials and their advisers when regulating enhancement technologies.

What Should Be Done?

Farrelly and Green have developed plausible models for intervention by the law in the area of enhancement technologies. Farrelly's "reasonable genetic intervention model" is intended to control governmental interventions in the reproductive freedom of parents. This model requires that reproductive freedom can be limited if (and it is implicit in Farrelly's [2002, 149–153; 2004b, 27–28] argument that he means *only* if):

1. The objective behind the measure that will limit reproductive freedom relates to concerns that are pressing and substantial in a free, democratic society.

and

2. (a) The means chosen are rationally connected to the objective.

    (b) The measure impairs reproductive freedom as little as possible.

    (c) There is proportionality between the effects of the measure and the objective that has been identified as of sufficient importance.

Farrelly contends that past eugenic programs, such as those pursued by Germany, the United States, and other nations in the earlier decades of the twentieth century, violated the first part of the model in making socially unnecessary government interventions. He suggests that we may be justified in intervening if our reasons are based on the same kinds of concerns that support reproductive freedom in the first place, such as concerns for autonomy, individual good, or equality. As elaborated by Farrelly, the idea of a rational connection requires demonstrable scientific evidence, a *need* for the intervention (for example, the intervention imposes conduct that parents would not engage in voluntarily in any event), and the likelihood that the intervention will be successful in its aims. In addition, there is an requirement to impair freedom as little as possible, and maintain proportionality between the aim of the intervention and the burden imposed (Farrelly 2002, 150–152).

Green's discussion is focused more specifically on the situation of children who are brought to birth following genetic interventions, though he also examines the issue of positional advantages, and, with reservations, he offers a guideline—his fourth of four—with a strong social dimension. His four guidelines are as follows:

1. Genetic interventions should always be aimed at what is reasonably in the child's best interests.

2. Genetic interventions should be almost as safe as natural reproduction.

3. We should avoid and discourage interventions that confer only positional advantage.

4. Genetic interventions should not reinforce or increase unjust inequality and discrimination, economic inequality, or racism. (Green 2007, 216–217)

It appears to me that each of these proposals contains some wisdom, although Green's four points can be read as a gloss on the first part of Farrelly's model—that is, they suggest a range of reasons why a free and democratic society might be justified in curtailing reproductive freedom: an intervention proposed by parents might not reasonably be regarded as in the child's interest; the technology might not be safe; the intervention may confer a merely positional advantage; or such interventions, perhaps cumulatively, may have certain adverse social impacts.

In what follows, I propose to draw on all these ideas, but with some different stresses. One difference is that I am mildly skeptical about any role in public policy for the concept of positional advantages (see chapter 3). It is difficult to find benefits that yield a positional advantage without also giving at least some other benefit. While there may be some extreme winner-take-all advantages that should be discouraged, a liberal society should hesitate to draw a regulatory line that rules out "ordinary" sorts of outstanding athleticism, say, without something more—such as risks to individual health and flourishing. Furthermore, as Buchanan (2011b, 48–50) has emphasized, many advantages, such as those in cognition, disease resistance, and span of life, may actually be more beneficial to each individual the more widely they are enjoyed by others.

Like Green, I also have reservations about a guideline that speaks loosely of reinforcing or increasing such things as "unjust inequality," which is a fiercely contested concept. What a socialist philosopher would consider a case of unjust inequality might strike a political libertarian such as Nozick as an acceptable and just outcome from voluntary transactions. For similar reasons, talk of economic inequality is far too broad—liberal democracies tolerate many things that have some tendency to produce economic inequality, while at the same time running a system of taxes and transfers to keep inequality under control. Even talk of increasing discrimination and racism is too broad. A liberal society does not attempt to suppress every possible action that might contribute, however indirectly, to racism or (unjust) discrimination, even though it may well conduct programs to ameliorate the impact of discriminatory thinking on people who belong to vulnerable groups, such as gay men, lesbians, women, African Americans, Australian Aborigines and Torres Strait Islanders, and others.

In chapter 4, I suggested that it is illiberal to demand that other people have homosexual children against their will. We should not accept that a homosexual man, say, is harmed in the legally relevant sense by the choices of remote individuals, whether they intend harm or not. Similar reasoning applies to the idea that a black couple might, for whatever reason, employ PGD or genetic engineering in an attempt to have a child with lighter skin. Even if we disapprove of the couple's motivation, it is illiberal to attempt to prohibit them from doing this on the ground that their action is indirectly harmful to remote individuals (perhaps because it somehow expresses or reinforces a belief that there is something wrong with dark skin). Agar (2004, 156–157) disagrees with this:

It is hard to imagine a successful fight against prejudice in the very society in which there is a widely exercised freedom on the part of parents to remove from their children the characteristics that would make them the object of prejudice.

I sympathize. Yet there are countervailing considerations. In an effort to allow individuals to pursue their own conceptions of the good, we often tolerate private behavior that is at odds with public policy and may have some adverse indirect effects on our policy objectives. For instance, it might be difficult to imagine a successful fight against homophobic prejudice in a society where people are free to accept and promulgate religious teachings that homosexuality is a sin—and many people actually exercise this freedom. Unfortunately, whatever its advantages in enabling people with many different ideas to live together peacefully, a society with liberal aspirations must to some extent tolerate the shortsighted (such as black couples who might choose to have lighter-skinned children) and even the intolerant (such as pious, or not so pious, homophobes).

What liberals must imagine is the likelihood that irrational prejudice will ultimately be easier to combat in a society where people with many different values are free to act in many different ways—verbally or otherwise—to influence the moral ethos. Against that backdrop, we should be reluctant to prohibit acts that have only an indirect and remote effect on the fight against prejudice. At the same time, we should exercise our own freedom by speaking up against the irrationality and cruelty of, to name two, homophobia and racism.

That said, liberal tolerance is not a suicide pact. Where there is a compelling case that certain behavior will (most likely) lead to intolerable social outcomes, and where it is unrealistic to resist this by noncoercive means such as criticism, persuasion, and personal example, there may be a case for prohibition. This confirms the thought that harms need not, strictly speaking, be *direct* for the state to be justified in exercising coercive power (Blackford 2012, 73–76). To be quite clear on this point, I am not thinking of so-called indirect discrimination, as when a firm has no specific policy against hiring women—but does have a policy against hiring workers of less than a certain height. In such cases, antidiscrimination legislation typically imposes an onus on the firm to demonstrate why the policy was independently and reasonably justifiable. The effects of personnel policies such as these are, in one sense, indirect; the impact on a particular individual is far less so. This is distinguishable from the more remote social impact of private activities that is generally permitted by the Millian harm principle.

To justify a legal prohibition, we need to identify some urgency or inevitability that goes beyond a mere tendency to cause harm by an indirect route. In the case of a dark-skinned couple's decision to have a lighter-skinned baby, we may well have reason to engage in moral criticism and persuade them, if we can, to act differently. As Mill (1974, 68) might have put it, there could be good reasons for remonstrating or reasoning with them, or persuading or entreating them, but it is not at all apparent in a situation like this that we should *compel* them.

It would be far too broad to propose that the state prevent all uses of enhancement technologies that have some negative impact on such things as economic and racial equality. What is required is a more convincing case that allowing such uses of technology will, by more or less unavoidable paths, lead to intolerably destructive or harsh social outcomes. In those circumstances, the state's concerns can be considered, in Farrelly's terminology, "pressing and substantial."

I take it, then, that we currently face a situation much like the following. We have powerful reasons to ensure that any research on enhancement technologies, particularly reproductive cloning and genetic engineering, is covered by strong and effective ethical standards with adequate teeth. This applies most obviously to any experiments that would involve radical modifications to human DNA such as by introducing sequences of genetic material from other species. A wide range of experiments would be irresponsible, though, if they involved bringing children to birth, perhaps with unwanted characteristics such as illnesses or disabilities.

There is also a case for discouraging some uses of technologies that are already possible or even foreseeable in the short term, such as those involving the use of PGD to select embryos with clearly detrimental characteristics. Nonetheless, I have argued that at this stage, it may *not* be best, all things considered, to enact legal prohibitions.

More exotic policy questions will arise when and if some of the problems of safety and efficacy are overcome, and enhancement technologies become more powerful. The case for the prohibition of a safe, effective reproductive cloning technology would be weak, though specific parents might still have sound reasons, in many or most circumstances, to avoid using the technology—given the risk that their children could face stigma or ostracism. There also appear to be legitimate concerns about the psychological effect on children if they discover they have human designers, and this could be particularly so if the design is thoroughgoing or involves direct attempts to design in personality traits. These concerns are not necessarily at a level that would justify regulatory prohibitions, but they mark out one area where vigilance might be required.

The most difficult questions for any policy approach to more advanced enhancement technologies are those that relate to (seemingly) beneficial interventions, especially if genetic engineering is used, potentially giving parents a wider palette of characteristics to choose from than offered by PGD. Here, the risks are likely to be indirect and long term, yet we can still imagine circumstances where the eventual impact would be intolerable.

A Policy Approach

Aggressive Regulation: For and Against

If truly powerful forms of human enhancement were available, and assuming that they were effective and safe, they could produce great benefits. At the same time, they could in principle pose long-term risks to society as a whole, threatening social instability or extreme inequality—or both. A number of points can be put to support aggressive deterrence of these particular risks (more than with other risks, such as those involved in allowing private schools).

First, advanced technologies of genetic choice might be relatively easy to suppress at this stage, compared with other technologies and activities. That is because such technologies as the use of genetic engineering to edit in or boost desirable traits do not yet exist. There are no existing markets to dismantle, and individual citizens have formed no emotional attachments to using such technology. Legal prohibitions could relate to the *development* of new technologies rather than the use of technologies that are already established and valued.

Second, the potential risk is radical, at least in principle. Though the social effects might be indirect and long term, some kinds of genetic engineering could strike in unprecedented ways at the perceived commonality among human beings or our natural responsiveness to one another. Alternatively, they could—again, at least in principle—create an unprecedented kind of gap between dominant and subordinate castes.

Third, although the nature of the longer-term risks can be described in a general way, it would be unreasonable to require a high standard of proof (including rigorously obtained, scientifically controlled empirical evidence) that particular social effects would eventually result. Unfortunately, such claims would be difficult to prove with scientific rigor, even if they were correct. The only way to be sure that the feared outcome would eventuate would be to wait until it actually did so!

Some considerations run the other way, however. Most obviously—though this is frequently neglected—the *full* range of values needs to be considered. Such values as the enlargement of human capacities, synergistic or network effects if enhancements became widespread, and the cultural products that could result should not be left out of the discussion. Second, we can ask for at least a prima facie case that some unwanted outcome is likely—enough of one to shift the burden of proof. This is very much lacking from prophets of a genetic aristocracy or a sharply differentiated caste society.

Moreover, we should always be reluctant to criminalize and stigmatize conduct that will cause only indirect and somewhat speculative harms. If an individual conducts research into enhancement technologies, with their obvious benefits, or adopts technological methods to enhance her children's genetic potential, this does not reveal a violent, dishonest, or otherwise-vicious character. Someone who does these things is likely to be motivated by their potential benefits, and is not thereby revealed as a person who is otherwise likely to be a danger to their society. This is an important consideration for lawmakers: otherwise-liberal societies could find themselves expanding the net of stigma and punishment, capturing people with dispositions of character that we do not ordinarily condemn, and who are unlikely to break other laws in any significant way. Other things being equal, that would be most unfortunate.

Ethical Standards and Principles for Regulation

With those considerations in mind, I propose a policy approach that does not attempt to demonize human enhancement or any form of it, or remove it forever from the boundaries of social acceptability. Indeed, this approach is open to human enhancement's possible benefits and welcoming them as far as possible.

For one, purely for safety reasons (in the sense defined in previous chapters), all affected jurisdictions will need to ensure that there are clear, appropriate, and effective ethical standards in place. These should apply to any activities in embryological research or reproductive medicine that could lead to the birth of human babies created by SCNT or modified by genetic engineering. Breach of these standards should have real consequences (but not the sort of severe criminal ones attaching to traditional crimes such as murder, rape, or armed robbery).

Issues of safety, which may change with time, do not point to the enactment of statutes designed to stigmatize SCNT or genetic engineering as inherently unacceptable. They do, however, suggest the need for the ongoing review and revision of applicable regulatory instruments to ensure that they operate as a clear, effective deterrent to irresponsible research that could lead to the birth of children with congenital deformities. At the same time, the aim should *not* be to deter practices that can be viewed as reasonably safe, based on evidence from previous steps taken within the jurisdiction or the experience in other jurisdictions. This area of policy should be flexible and adaptable, since the state of the art in reproductive medicine does advance—though not as rapidly as alarmists usually indicate.

While this provides a basic level of regulation to meet safety concerns, lawmakers may need to enact some additional provisions. They should be alert to the possibility of harmful outcomes even from the safe and effective use of enhancement technologies. The harms might be relatively direct (such as the psychological impact on children), or long term and indirect, such as pressures toward a more hierarchical society. But lawmakers in a liberal democracy should display a degree of restraint, possibly expressed in a set of principles for policy development that could be disseminated widely, and relied on to combat knee-jerk reactions to public fears or pressure from lobby groups.

Below I set out four principles, each with its numbered heading followed by some explanation and elaboration. I then add a final comment on the often-forgotten merits of inaction.

Principle 1: *The Need for a Compelling Case*

The first principle roughly corresponds to Farrelly's principle that there must be concerns that are pressing and substantial. Such concerns will usually involve significant, direct, and wrongful harm to others' secular interests. The Millian harm principle may not be absolute, but any exceptions to it must be consistent with the liberal values that it protects. As I've expressed the point elsewhere:

If coercion can be justified in respect of some indirect harms, it is to the extent that they resemble more direct ones in the need for an urgent response; if it can be justified in respect of some kinds of "mere offense," it is to the extent that the impact of offense merges with that of unequivocal harm; if it can be justified in respect of some self-inflicted harms, it is largely to the extent that we have good reason not to trust our own judgment in areas requiring sophisticated technical competence. (Blackford 2012, 76)

Where the argument for regulation involves a fear of social disintegration or perhaps the specter of a caste society with harsh outcomes for a subordinated class, it might be unreasonable to demand empirical evidence presented to a high scientific standard. Nonetheless, something rationally compelling needs to be said as to why the postulated technology and practices associated with it could be expected to become widespread, why they might reasonably be expected to generate unprecedented social risks, and why they could not be accommodated by modern-day liberal democracies. After all, large claims require convincing support. Some risks admittedly may be so great that even a low probability might justify avoiding them at all costs—risks of human extinction, for example—but even then, the claim must be backed up by something realistic, not merely fanciful. Otherwise, we are captives to any and all claims that a wide variety of innovations could, by one path or another, lead us to our doom.

We should not allow opponents of new technologies or practices to get away with irresponsible fearmongering about horrible, but distant and speculative, scenarios. Instead, there are pertinent questions to be asked. Just what evil is expected to eventuate? How will it come about, and by what mechanism? And what existing or likely circumstances might prevent the worst scenarios?

It should not be good enough to claim, as Mehlman (2003, 189) does, that (say) the genetic engineering of the human germ line threatens "the destruction of democracy and the enslavement of the species." If legislators believe that such an outrageous result is likely to eventuate from certain innovations, let them at least postulate a plausible mechanism. Further, they should explain why the innovations could not be accommodated by other means, such as diligent efforts to expand access to new technology, combined, if needed, with laws aimed at protecting genetic information, deterring unreasonable forms of genetic discrimination, and ensuring that possession of enhanced traits does not confer greater political or legal rights and liberties.

Principle 2: The Need for Narrow Drafting
The state should avoid suppressing technologies and associated practices where this is not specifically justified, and especially when doing so might entail its citizens' forgoing important benefits. Accordingly, any legal prohibitions should be framed as narrowly as possible, their drafting should be construed strictly by administrators and courts (and should explicitly invite this), educational campaigns should have narrow objectives, and so on. The aim should always be to preserve as much individual liberty as reasonably possible.

Even if some prohibitions are put in place, exceptions may be required. For example, a ban on the direct engineering of personality traits might leave room for "correcting" a propensity to psychopathy. A ban on the transgenic genetic engineering of human embryos might allow a specific exception for the insertion of a DNA sequence that is known to be safe and beneficial for the individual, and reasonably believed to be of long-term social benefit. The case for such an exception might be compelling in a situation such as Harris (1998, 20–21) discusses, where transgenic engineering is used for limited and essentially preventive purposes, such as if a nonhuman gene could confer resistance to certain environmental pollutants.

Thus Mehlman goes down the wrong path when he maintains that we must prevent all research, public and private, on germ line enhancement along with all patents on germ line genetic engineering. This must include research on animals as well as humans, he asserts, because it could theoretically be used to create a rival intelligent species. In Mehlman's view, sweeping prohibitions could be supported by provisions to revoke licenses for fertility clinics and doctors, disentitle hospitals and clinics from carrying out relevant research, and bar public funding and approval of otherwise-legal products for companies and researchers. As a last resort, genetically enhanced individuals could be sterilized. Ironically, this vision of near-totalitarian control is intended to avoid what Mehlman (2003, 184–186) often refers to as a "nightmarish future." Contrary to his approach, we need drafting that targets realistic dangers, not a regime of extensive prohibitions based on speculation and fear.

Consider Mehlman's wish to avoid the creation of intelligent species that could rival Homo sapiens. Strictly speaking, any such proposal lies beyond the technologies considered in this study, which relate to *human* genetics; proposals for "uplifting" nonhuman animals to higher levels of cognitive capacity or creating entirely new intelligent species merit separate, perhaps lengthy, consideration. Yet the issue of what might be called nonhuman enhancement is worth a brief comment at this point because it can be used to highlight the flaw in Mehlman's approach.

Though some contemporary thinkers take a different view (e.g., Hughes 2004, 91–94), I suggest that Mehlman has good reason to be concerned about creating a new species with human-level intelligence. The problem does not relate to any intrinsic immorality in crossing species barriers or creating transgenic organisms. Like Pence (2004, 142–161), I suggest that we try to consider the actual characteristics of any proposed animal—focusing on the individual merits of each case. In instances involving uplifted animals, the members of a new, transgenic species would (I assume) be the result of modifications to the DNA of existing nonhuman species. Even those species that are genetically closest to us come with evolved behavioral propensities that are not human and are likely to be quite different from our own. In particular, they are not equipped—or certainly not to the same degree—with our natural capacity for affective communication with fellow human beings. Hence, any uplifted, hyperintelligent animals might be poorly adapted to living with us as fellow citizens. If let loose in human society, they might be dangerous to us, and we to them.

There is doubtless more to be said on both sides of this specific issue, but I believe that a strong case can be made against such a project, at least until far more is known about the psychological mechanisms underlying human sociability. In our current state of knowledge, creating uplifted nonhumans with human-level cognitive capacities might be dangerous to all affected. Moreover, it would not be satisfying the existing preferences or interests of existing sentient beings; it would instead bring about new beings with altogether-new preferences and interests—beings with vulnerabilities and yearnings that current chimpanzees and other nonhumans do not have.

All that conceded, Mehlman overreaches in calling for a wide ban on research that involves the genetic engineering of nonhuman animals and modification of their phenotypic traits. If there is ever any serious prospect of creating nonhuman rivals to our own species, prohibition may indeed be an appropriate response. When and if necessary, we could draft narrow legislation aimed specifically at forbidding attempts to bring hyperintelligent nonhumans to birth. But as long as such animals do not develop as far as being born and interacting with human beings, there is no potential for any significant harm to come of it. There is no slippery slope here, with a horrible outcome at the bottom, so long as we understand clearly what it is that we wish to prevent, and why.

To repeat, the problem is not that creating hyperintelligent or uplifted animals breaches a taboo against genetic engineering or crossing species barriers—a taboo that would be encountered and set aside at an early point on the supposed slope (in fact, we are already well past this point). Rather, the concern is based, quite narrowly, on the dangers from (and to) new sorts of animals that would fit a specific description. If it ever appears necessary to draft regulatory instruments to forbid this, they can and should be narrow as well as specific in their wording.

Principle 3: The Need for Flexibility

Since any prohibitions should be framed in narrow terms, they will not prevent continued research on related technologies. For example, a ban on some radical uses of enhancement technologies, such as attempts to create a transgenically engineered child with gills, an eel's electric sense, or (more simply) fluorescent skin, would not prevent less radical uses, such as PGD for sex selection or choosing children with the potential for long, healthy lives. Nor would it prevent the use of SCNT for reproductive cloning if some breakthrough enabled this to be done safely.

If some of the technical barriers to safe, effective human enhancement are eventually overcome and some citizens begin to use the new technologies, governments and other parties will be able to monitor the benefits obtained, along with both desirable and undesirable impacts. Undesirable impacts could relate, for instance, to adverse psychological effects on children, who might suffer from stigma or confusion. Efforts would need to be made to compile extensive and reliable data rather than relying on data from a self-selecting group of unhappy individuals. Accurate data might eventually provide evidence to justify tougher restrictions, but it might also tend to show the opposite. It might justify relaxing or repealing precautionary regulations.

The provisional or tentative nature of any legislative prohibitions suggests that any restrictions on enhancement technologies should be in a form that is conducive to easy amendment or repeal. For example, incorporation in subordinate legislation or administrative codes of practice would be more appropriate than inclusion in a major penal statute, such as a jurisdiction's criminal code or crimes act. As Stock (2002, 209) suggests in his discussion of "germinal choice technologies," it is better to use narrow administrative guidelines, as opposed to sweeping legislation, in order to retain our capacity to make refinements as events unfold.

Other means might also be used to ensure that benefits are not lost permanently. Any legal prohibitions could be drafted with sunset clauses, as with some provisions of Australia's 2002 Prohibition of Human Cloning Act[1] in its original form. Limited licenses could sometimes be used to test an innovation's social effects on a relatively small scale, as proposed by Harris (2007, 148–149) with embryonic sex selection.[2] And all possible steps could be taken to avoid creating (or reinforcing) the social or moral stigma that would make the later repeal of prohibitions politically difficult.

This leads to my fourth and final principle.

Principle 4: The Need to Avoid Moral Stigma

Any prohibitions should be framed and given political support in ways that are calculated to avoid stigmatizing the prohibited practices as inherently immoral. Moralistic posturing by the state should be avoided. Rather than the assignment of moral guilt or an attempt to lead the electorate into the acceptance of new moral norms, the emphasis should be on society's practical need to take care as it accommodates new technological practices.

This goal can also be furthered in a variety of ways. Importantly, any political rhetoric that is used to support prohibitions should be measured, perhaps even regretful. Such actions as the imposition of crushing terms of imprisonment should be avoided, and (as already suggested for a different reason) any formal prohibitions should *not* be placed in a legislative instrument, such as a statutory criminal code or crimes act, that also covers such core components of criminal law as murder, rape, theft, and fraud. The only obvious exceptions would be for crimes that involve truly malign or sadistic choices of detrimental traits, using PGD or genetic engineering—should it ever appear necessary, all things considered, to create such statutory crimes.

Prohibitions on certain areas of research should be enforced in ways that are relatively open to review and withdrawal, such as the imposition of civil or regulatory penalties, the nonavailability of certain intellectual property protections, and steps to vet the publication of certain kinds of research. For example, interested governments could pressure the publishers of scientific journals not to accept submissions that report experiments aimed at bringing to birth human beings with transgenic DNA sequences. These steps would be in addition to funding penalties for organizations dependent on support from public sources. The precise form of any regulation could be decided after widespread consultation focused on practicalities and would take into account the views of the relevant research communities.

The Forgotten Merits of Inaction

Since Dolly was born in 1996, and announced to the world in early 1997, it has become increasingly clear that there was no need for specific laws prohibiting human reproductive cloning. Given the difficulties even in cloning nonhuman primates, the relative lack of scientific eagerness to produce the first cloned human child, and the low level of demand, all the energy, time, and money put into enacting legislative prohibitions has been a waste of public resources. Though the extent of the barriers to human reproductive cloning was not fully predictable in, say, 1997, it has been well known for years now; yet efforts continue to be made to pursue an illiberal policy that would not only prohibit reproductive cloning while it remains unsafe but also attempt to stigmatize it for all time.

Much the same applies to other uses of enhancement technologies. As one illustration, there seems to be no compelling case to ban sex selection. As another, there is a good case, in current circumstances, to *tolerate* the selection of embryos for such characteristics as deafness, notwithstanding recent and current debates over the issue. There also appears to be no current need for legislation to protect embryos that are intended to be brought to birth after attempts at detrimental genetic engineering. Why enact a law against something that almost no parent would seek to do in any event?

If the present hurdles to more powerful technologies are overcome by some unforeseeable means, I do see two relatively clear and natural-seeming lines that we should be cautious about crossing: attempts to control our children's personality traits directly by genetic engineering; and efforts at radical genetic engineering by introducing new nucleotide variations or entire strands of transgenic DNA into the genotypes of human embryos. The former creates a plausible, though speculative, risk of making children feel like puppets—even if feeling like that would not be entirely rational—while the latter would not only raise serious safety concerns but also open up the (again speculative) possibilities of creating genetic gods and monsters.

And yet there is no real necessity for legislation about crossing these lines—at least not at the present time. Caution does not mean absolute prohibition, and these uses of genetic technologies are still too distant, separated from us by substantial technical difficulties. Doubtless it is worth identifying which lines are potentially important, and this can be factored into future policy consideration if the technology advances in unexpected ways. Other than that, there seems to be relatively little that a liberal state should currently be doing to prohibit the more exotic uses of enhancement technologies. Government officials may feel the need to act or be *seen* as acting, but there can be merit and wisdom, and a form of mastery, in well-chosen political inaction. Its merit would be even greater if lawmakers in liberal democracies were more willing to provide their electorates with careful, but clear and firm explanations as to why legislative restraint is often for the best.

Conclusion

If powerful forms of human enhancement technology ever become possible, they will bring with them both benefits and threats, although careful scrutiny of the threats suggests that many have been exaggerated. The innovations that could in principle produce the most dramatic alterations to human capacities are those involving radical alterations to the human genome as opposed to tinkering with its natural variations. The barriers to achieving this level of technology—barriers imposed by issues of efficacy and safety—are, as Allhoff (2005, 43) states, currently tremendous. Even if some particular alterations prove to be easy after further investigation, it is doubtful whether many parents would seek truly inhuman or super-human characteristics for their children. Seen in that light, much of the opposition to enhancement technologies is irresponsible fearmongering.

Nonetheless, some advances, perhaps even breakthroughs, can be expected. If certain prospects do become more realistic, there will be further pressures to prohibit or regulate, and some of those pressures should be resisted. In response to the sorry retreat from liberal tolerance that has characterized much of the debate that followed Dolly, I have proposed four principles to guide lawmakers: the need for a compelling case; the need for narrow drafting; the need for flexibility; and the need to avoid moral stigma. If these principles were followed, a genuinely liberal approach might emerge, and in many cases the best policy prescription might turn out to be one of legislative restraint.

The principles that I've suggested are not meant to set an easy test. They should be applied scrupulously and honestly, against a background of ethical standards relating to such mundane issues—for bioethicists—as safety, adequately informed consent, nondiscrimination, and the privacy of personal information.

It is certainly important to identify the genuine dangers from new technologies, wherever we find them lurking in the penumbra of technological progress. The technologies of genetic choice will doubtless bring their share of problems, but they will also bring benefits. Indeed, my proposed principles justify the prohibition of little (if anything) in the way of actual, or realistically imminent, technologies or uses. Or so it appears to me. To take this a step further, I'm convinced that the crisis we currently face is *not* the coming of Frankensteinian or apocalyptic technologies that must be controlled as a matter of urgency. Rather, there is a crisis for liberal tolerance.

Consider the frequent policy distortions, emotive overreactions of many commentators, and political overreactions of many legislatures; consider, too, the unashamed antiliberalism that so often characterizes the human enhancement debate and provides a precedent for other debates to come. Faced with all this, we can dissent. Whether or not our dissent is heeded in the short-term political process, we can maintain it with conviction and strong argument. We can maintain the candles of liberty and reason in what has become a dark area of public policy (Pence 2004, 181).

Throughout this volume, I have focused on a specific set of emerging technologies that offer us genetic choices. There are, however, wider implications. The future will bring new political issues of many kinds. Some innovations will provoke irrational fear and anger, as did the cloning of Dolly and the prospect of reproductive cloning for human beings. Next time such an issue emerges, it will again grant our lawmakers an opportunity to show their credentials as successors to Locke and Mill. May they rise to the challenge.

Appendix: The Therapy/Enhancement Boundary

The purpose of this appendix is to address some conceptual and terminological difficulties that bedevil wider debates about enhancement, and particularly a supposed distinction or boundary between enhancement and therapeutic interventions. I will show something of how problematic this is.

Among others, the authors of *From Chance to Choice* have already made the most crucial points: any boundary that can be drawn between therapy and enhancement (in the sense under discussion in this appendix) fails to correspond with a moral boundary between the obligatory and nonobligatory. Nor does it correspond with a boundary between the morally permissible and impermissible. Furthermore, we should acknowledge that limited resources prevent treatment of all diseases and impairments, that some interventions that go beyond therapy might be socially desirable and worth funding, and (though this seems more problematic) that there could even be odd cases where a specific kind of therapeutic intervention is best regarded as morally impermissible (Buchanan et al. 2000, 120–121, 152–154).

I do not propose to argue that a therapy/enhancement boundary can *never* be identified, but it is not a reliable, workable boundary. And even if we could identify it with scientific objectivity across a wide range of cases, it might deliver counterintuitive or confusing results. For example, it offers dubious guidance when it comes to psychological dispositions and age-related forms of decline.

Approaches to the Therapy/Enhancement Distinction

As Agar (2004, 79–80; cf. Kitcher 1996, 208–209) explains, the relevant philosophical and bioethical literature reveals two main approaches to defining enhancement in opposition to such concepts as therapy. Using

a social constructivist approach, diseases are states that are viewed nega-
tively by society. Just which interventions we should regard as therapeutic
is thus, at least to some extent, a matter of social choice and not simply
given to us by nature. In this spirit, Pence suggests that our concept of
disease is partly evaluative, based on when we see a particular abnormal trait as bad. According to Pence's model, we have a core concept
of disease with expanding circles beyond it, where application of the
concept is increasingly controversial. Disease and dysfunction shade into
normal functioning, which shades into ideal functioning, which is itself
altered over time. Human life expectancy, for instance, has altered greatly
from the norm of about thirty-five years in ancient Greece (Pence 2000,
109–110).

Based on an objectivist account, by contrast, the definition of disease is
independent of our attitudes and relates to some part of a person's body
failing to perform its biological function. One influential version of the
objectivist account can be found in the work of Norman Daniels (1986,
26), who employs the concept of "species-typical normal functioning."
This is also used in the more recent *From Chance to Choice*, of which
Daniels is a coauthor.

According to Daniels (ibid.), the idea of health can be elaborated as
an absence of diseases, deformities, and impairments (including those
resulting from injury) that "are *deviations from the natural functional
organization of a typical member of a species.*" In the case of human beings, health must also involve those aspects of our functional organization that enable our pursuit of "biological goals as social animals." Hence,
for Daniels (ibid., 29), normal human functioning includes such things as
language use, knowledge acquisition, and a capacity for social cooperation—and it extends to mental health.

Human health care needs, viewed through the lens of an objectivist account, are those necessary to maintain or achieve functioning at an objectively healthy level. Daniels claims that harmful deviations from this are
regrettable not only for broad utilitarian reasons (their tendency to create
pain or frustrate individual preferences) but, more importantly, because
they can limit people's abilities to carry out what would otherwise be
reasonable life plans. He suggests that the reasonableness of life plans can
be assessed in relation to normal opportunities in the relevant society as
well as an individual's particular talents and skills (ibid., 27–28, 33–35).
Daniels maintains that the state should provide health care not only to
reduce suffering but also because of the positive effect that meeting health
care needs has on individuals' life opportunities.

If we follow Daniels this far, we can use the word therapy to refer to the restoration of normal human functioning—which provides the baseline for what is regarded as health. Within such a model, enhancement involves interventions "to improve human form or functioning beyond what is necessary to sustain or restore good health" (Juengst 2000, 29). This formulation by Juengst usefully refers to form as well as functioning, and therefore covers interventions for aesthetic purposes, such as cosmetic surgery—or indeed, genetic modification—aimed at increasing physical beauty or sexual attractiveness. In his recent book *Better Than Human*, Buchanan (2011a, 5) provides a similar definition: "An *enhancement* is an intervention—a human action of any kind—that improves some capacity (or characteristic) that human beings ordinarily have or, more radically, produces a new one." Unfortunately, the seemingly clear distinction between therapy and enhancement is beset by ambiguities and problems. Before I come to these, however, I wish to emphasize one powerful consideration in its favor.

Our various organs and systems of organs—the eye, the liver, the neurological, pulmonary, and cardiovascular systems, and so on—were not designed by intelligence. Yet they are adaptively complex structures. They and their components can be construed metaphorically as having been engineered by natural selection to function in particular, rather precise ways. Thus, it makes sense to think of the human body as an engineered structure of structures, and even to speak of a correct way for it and all its complex components to function—so long as we remain aware that this is metaphorical (Kitcher 1996, 210–212). To this extent, biomedical science can increasingly discover and describe normal human functioning, and identify deviations from it.

Accordingly, Agar (2004, 79–80) is correct that constructivists make the nature of disease too dependent on current social reality. There does seem to be a biological line that can be drawn between the "correct" functioning of the body (with its organs, systems of organs, and so forth) and harmful departures. As we will see, though, that line does not necessarily fall where theorists such as Daniels would place it.

Some Problems for the Distinction

Is It All Enhancement?
An initial objection to the distinction between therapy and enhancement might be that therapy is itself enhancing in obvious ways; therapy for a disease, injury, or disability, for instance, is intended to produce an

improvement in the patient's bodily functioning (Savulescu, Sandberg, and Kahane 2011, 6–15). So it's *all* enhancement!

Still, therapy can be distinguished as helping individuals in a specific way: by bringing an individual's bodily functioning closer to the species-typical norm (Parens 2000b, 5; President's Council on Bioethics 2003, 14–15). Hence this objection is not successful, because we can easily think of real and imaginary cases where someone attempts to produce an improvement that seems to be more than therapeutic—perhaps by taking an individual whose bodily functioning is already normal, and thus who is not sick, injured, or disabled, and attempting to bring their functioning to an augmented level. Even if we think, "It's *all* enhancement!" we are likely to distinguish enhancement that takes the form of treating disease from enhancement that increases the capacities of people who are not suffering disease (whether the increase is within or beyond the typical range for our species) (Savulescu, Sandberg, and Kahane 2011, 8).

At this stage, I submit, the therapy/enhancement distinction seems intuitive, and it might be prima facie satisfactory for some decisions relating to, say, health care provision or health insurance coverage. I am sympathetic to the case that at least the primary justification for health services is to provide reasonably effective treatment for diseases and impairments that have a negative, nontrivial impact on individuals in the society concerned (Buchanan et al. 2000, 121–124). This idea tracks many ordinary concepts of fairness, is reasonably effective and affordable to administer, lends itself to actuarial calculations, can be (somewhat) objective, can avoid certain moral hazards, and places a responsibility on people for their own tastes if they seek more than good health (ibid., 112–114, 142–144). Many exceptions to such a policy position might be justified, perhaps on a case-by-case or class-by-class basis; the human body's natural functional organization, however, appears to have at least some biological, social, and political relevance.

Nonetheless, there are problems with the idea of an objective boundary between therapy and enhancement.

Too Many Categories?

For a start, absurdity can result if we attempt to classify every kind of technological intervention in the form or functioning of the human body, no matter how trivial, as either therapy or enhancement. Most obviously, some interventions intuitively might be neither of these if they are actually detrimental. In addition, the words therapy and enhancement are not useful to characterize relatively trivial interventions, such as tattooing and hairstyling, even though these might be said to improve the human form (at least by someone's subjective standards) in ways that are not needed for health. Surely we at least need to distinguish between embellishments that are, as it were, skin deep, and those that are more intrusive and far reaching in their effects.

Trivial beauty treatments should remind us that even long-accepted activities, such as the use of tools and chemical products to cut and style hair, can have dramatic effects on an individual's appearance and presentation, and that even *this* may not be entirely unproblematic if we are deeply concerned about issues of fairness. An awareness of skin-deep embellishments may remind us of the pervasiveness of social, economic, and sexual competition in human societies, the universal employment of multifarious technologies to gain various advantages, and the many related moral questions, large and small, that might arise.

Beyond noting that point, however, nothing is gained by thinking of haircuts and tattoos as being human enhancements. Let us assume that the whole debate is about more intrusive technological interventions. Even so, there are serious questions about, for example, which more substantial interventions should be provided by a health care system, which should be covered by insurance, and which might be prohibited.

A more serious set of problems relates to how we should classify disease prevention. Before elaborating on this, I need to make it clear that Daniels's own concept of health care is wider than that of therapy or treatment. He describes a health care system as involving "a diverse set of institutions which have a major impact on the level and distribution of welfare" (Daniels 1986, ix). Indeed, his underlying rationale for health care supports many activities in addition to therapy for the sick and injured. If the purpose of health care is to ensure normal human functioning as far as possible, in order to facilitate individuals' ability to make and carry out "reasonable" life plans—or for some broader set of consequentialist reasons—then *prevention* of rationally feared things such as disease and injury is a legitimate part of health care. Accordingly, health care can be understood as an umbrella concept within Daniels's analysis: it includes the provision of therapy, but also vaccinations, occupational health and safety regulations, product warnings, and such familiar courses of action as assisting someone to follow a sound diet and engage in regular exercise. It also includes educational activities relating to any or all of these.

We should also accept that in many practical settings, health care includes activities that are more palliative than therapeutic in nature. In fact, as Daniels (ibid., 12–13) points out, medical practice was largely confined to palliative activities until fairly recent times, as prescientific medicine had only limited therapeutic and preventive efficacy. Health care can also include social support for those whose functioning is impaired as well as such methods of ameliorative intervention as the provision of surgical prostheses or wheelchairs.

A difficult issue arises with preventive measures like vaccination. Should this be regarded as a kind of preemptive therapy? Alternatively, is it enhancement, since it gives a new ability to individuals who are not already sick, and whose bodies may be functioning quite normally at the time? Once vaccinated, an individual is expected to possess a capacity beyond what is typical for human beings who have not undergone the technological intervention. The whole point is to grant additional resistance to harmful infection by particular microscopic life-forms. Based on such reasoning, Harris (2007, 21) considers vaccination to be an enhancement technology, although one that has encountered little opposition.[1]

Parens (2000b, 5) suggests that prevention could be thought of as a separate category in addition to therapy and enhancement, but I submit that the crucial point here is a practical one. Familiar medical interventions for the purpose of prevention—such as vaccinations—are intended to head off challenges to the body's normal functioning. For that reason, they are interventions that are closely related to therapy—that is, the medical treatment of those who are already sick, injured, or otherwise affected by impairment of functioning.

Importantly, vaccination does not challenge the concept of normal functioning. Vaccination does not improve just *any* aspect of the body's functioning; rather, it is aimed at the precise biological systems that confer resistance to disease. This highlights the fact that vaccination has a health care goal in the sense explored by Daniels. If the word enhancement has pejorative nuances, we may be motivated to withhold it from technological interventions that improve the body's resistance to agents or events that would otherwise impair normal human functioning.

Such an approach helps to explain Glannon's (2001, 95) positive attitude to the genetic manipulation of the immune system to increase resistance to disease. Though his official position is one of opposition to what he regards as genetic enhancement, Glannon (ibid., 96) does not wish to include genetically based preventive interventions. Indeed, he proposes excluding preventive medicine altogether from the concept of enhancement; for instance, he would exclude steps taken to increase bone density in postmenopausal women to resist osteoporosis.

This is not obviously implausible. Accordingly, examples relating to prevention do not seem to show that the therapy/enhancement distinction is untenable, though they do add a layer of complexity. We need to distinguish between therapy and the broader category of health care, and between medical interventions designed to improve resistance to disease and those that may be designed to improve various other capacities or traits. More complexity is added, however, by the theoretical prospect of achieving essentially preventive goals by using quite radical methods. As Harris (1998, 20–21) remarks, even transgenic genetic engineering might be used for such purposes if, say, the insertion of a nonhuman DNA sequence into a human genome conferred resistance to certain environmental pollutants.

If we are going to classify all interventions that aim to prevent disease as therapeutic or at least as health care initiatives that are akin to therapy, we ought to note that therapy and health care could in principle include some quite radical uses of genetic technologies.

Another problem is that some familiar kinds of intervention employ medical knowledge and techniques, are far from trivial, appear to be socially justified, and yet do not fall neatly into either therapy or prevention. Consider abortions. Some are in fact performed for the sake of a woman's physical or mental health, or even to save her life. These can easily be regarded as therapeutic, but they are far from being the only ones that are routinely performed. Or consider a woman who is using the contraceptive pill for its routine purpose of avoiding pregnancy. This is not therapy, as understood in the discussion so far, because there need be nothing unhealthy or impaired about the way her body is functioning— fertility is not, for example, a disease or injury. Nor, for the same reason, is contraception a case of preventing disease or injury; what it prevents is not a harmful departure from normal human functioning but instead part of what the (female) body is "engineered" to do.

I'm not suggesting that we should condemn women who use contraception or decide to have abortions, or governments that subsidize them in either activity. Rather, medical interventions and medical science can give people things that they value—in this case, including a wider, more equal range of opportunities for women—without falling neatly into the categories of therapy and prevention. As Daniels himself acknowledges, medical interventions can further social goals beyond those associated with preventing, treating, or otherwise ameliorating impairments of normal human functioning. Thus, permitting or even funding abortions might be morally acceptable as well as socially advantageous, even if the abortions concerned have no therapeutic function (Daniels 1986, 31–32).

Buchanan and his coauthors (2000, 120–121, 135) of *From Chance to Choice* appear to regard nontherapeutic abortions as enhancements, albeit enhancements that are socially justified. Yet no one's capacities or other favored traits are improved to a degree that goes "beyond therapy" when a fetus is aborted. The form and functioning of the patient's body are not boosted in any sense. Instead, if all goes well, her body will soon return to something like its condition before the pregnancy. In those circumstances, there is something odd about classifying abortions as enhancements, within a scheme that contrasts enhancement with therapy. It is better to acknowledge that many interventions, including justified ones, do not fit well within such a scheme.

The contraceptive pill is a rather different case, and perhaps it can be regarded as an enhancement in the relevant sense. Admittedly, the temporary suppression of the body's reproductive abilities is (perhaps) not exactly what we think of as an enhancement (Kamm 2009, 107–108). But consider all the following. First, the biological functioning of the woman concerned need not be impaired in any way before she commences taking the pill, so it is not being used as therapy. Second, while the immediate effect is to suppress something that the body is capable of doing, the larger purpose is to enable the woman to take ongoing control of her fertility. That does sound like a boost in her abilities. Within the constraints of the technology and some unfortunate margin of error, she is given a new power to turn her fertility on and off at her own convenience.

So perhaps we should regard use of the contraceptive pill as falling on the enhancement side of a therapy/enhancement boundary. All the same, such examples make it clear that not all desirable interventions can easily be categorized within this scheme. There is something procrustean about it.

Deeper Problems

Bioengineering Accounts and Their Limits
But this only scratches the surface. Recall that I offered some support for an objectivist approach to the therapy/enhancement distinction by invoking a bioengineering account of the body as a system of adaptively complex, interacting subsystems and structures. While this system of systems may be imperfect in various ways, biomedical science can increasingly describe its operations and identify departures from how it is engineered (or "engineered") to operate.

This need not line up well with a concept of *statistically* normal functioning. To take the example of height, somebody might be short or tall compared to the mean, but there might be nothing malfunctional about any of her organs, organ systems, and so on, from a bioengineering viewpoint. According to those standards, she might be perfectly healthy. At the same time, there could be many situations where her height is such a social disadvantage that it seems tempting to describe it as falling outside of normal human functioning and contemplate whether she is entitled to assistance from the health care system in the jurisdiction where she lives.

Fukuyama deals with what he sees as another such illustration—attention-deficit/hyperactivity disorder, which, he argues, may not be a disease at all but simply a matter of some people being at the tail of a normal distribution of psychological traits to such an extent that it interferes with their social functioning. He employs this case to make the point that it is sometimes easier to draw lines that are useful to regulators than to make distinctions that are beyond all theoretical challenge. In practice, he says, we can decide who gets Ritalin and who does not (Fukuyama 2002, 210–211). If Fukuyama's account of attention-deficit/hyperactivity disorder is essentially correct, he may also be right that this is an example of a practical decision that does not depend on any strict bioengineer's concept of species-normal functioning.

Note, however, that if we accepted Fukuyama's understanding of attention-deficit/hyperactivity disorder *and* relied on a bioengineering account of normal human functioning, we would conclude that all uses of Ritalin to "treat" this "disorder" are cases of enhancement. The lesson in all this is that our health care decisions ought to take into account the issues of social advantage and disadvantage raised by constructivists. These may lead us to treat people who, from the bioengineering perspective, are actually healthy.

There is more to say about why a bioengineering perspective can sometimes mislead us. Some departures from the human body's identifiable bioengineering may be too trivial to warrant any intervention (Buchanan et al. 2000, 117–119) or significant only in certain social contexts (ibid., 79). As regards the latter, Buchanan and his colleagues offer a striking example from life in a preliterate hunter-gather society, where one tribal hunter might have a neurological condition that could potentially impair his development of reading skills—which are, of course, irrelevant to his way of life. If there is no detriment to more relevant skills, such as hand-eye coordination, general motor capacities, and oral communication, the individual may not, for any practical purposes, suffer poor health.

Agar points out that there could even be biological malfunctions that actually *promote* survival—at least in a particular environment. For instance, a genetic mutation might interfere with the metabolism of nicotine and make the affected individual less likely to become a smoker (Agar 2004, 81). "Treating" this would (I take it) be a mistake. Thus, departure from normal human functioning—as viewed from the bioengineering perspective—is not disadvantageous in all conceivable circumstances.

Conversely, some individuals may experience considerable disadvantage and psychological suffering even though, from the bioengineering perspective, there is nothing wrong with them. As we've seen, diminutive stature might be one example. Harris offers a more science-fictional one. If there were further depletions of the ozone layer, and in consequence white people became especially vulnerable to skin cancer, it would be rational to think of white people as relatively disabled, notwithstanding that white skin is not species abnormal. This underlines his view that what really matter in medical practice and health care are conditions that it would be negligent not to fix if we could—something that is determined in part by technological and other circumstances (Harris 2007, 44–45, 92–93).

Wise decisions about whether any intervention should be attempted might often be based, quite legitimately, on relatively imprecise folk wisdom about what is painful, debilitating, or restricting. Detailed scientific knowledge of the functioning of organic systems also might be needed to ensure that interventions are actually effective.

Cognitive Capacities and Psychological Dispositions
Moreover, the bioengineering approach may be less than impressive when we turn to considerations of cognitive capacities, psychological dispositions (including sexual orientation), and mental health. In *What Sort of People Should There Be?* Glover (1984, 53) observes, "With emotional states or intellectual functioning, there is an element of convention in where the boundaries of normality are drawn." In his more recent *Choosing Children*, he discusses the "messiness" of any boundary between normal functioning and disability, including the problem that merely statistical differences will not work (Glover 2006, 12–13).

Emotional states and intellectual functioning may be reducible to or supervenient upon the functioning of the embodied, acculturated brain and neurological system. But when, exactly, is it possible to claim that an individual's neurological functioning has gone wrong? If the brain fails to perform its function of regulating the various subsystems of the body, such as respiration and heart rate, that should doubtless count as an adverse departure from species-typical functional organization. When we are confronted by variations in cognitive skills, though—such as whatever skills are measured by IQ scores—we are back to a statistical range. The same applies to the vast diversity of human emotional responses, dispositions, and attitudes. In most cases, there is no obvious bioengineering answer to what counts as the correct, nonpathological functioning of these things.

In respect to psychiatric conditions, the authors of *From Chance to Choice* appear to have a touching faith in the ability of psychiatry to produce objectively acceptable standards; they regard manuals such as various iterations of the *Diagnostic and Statistical Manual of Mental Disorders* as having biological validity for producing diagnoses of genuine mental diseases (see Buchanan et al. 2000, 142–143), and state categorically that homosexuality is not a disease (ibid., 122), as if this were simply a biological fact. In some cases, faith in psychiatric manuals certainly may be justified; departures from species-normal neurological functioning may underlie certain kinds of delusional or obsessive behavior, for example. But is the position really so clear, one way or the other, with the phenomenon of sexual orientation? This merits some further thought.

In 1973, the American Psychiatric Association acted to remove homosexuality from its *Diagnostic and Statistical Manual of Mental Disorders*, essentially because homosexuals were no more likely to be impaired (by other standards) than anyone else. For what it may be worth, I totally approve of this decision, although the classification of homosexuality as a disease was perhaps better, under the circumstances then prevailing, than classification as a crime (Kitcher 1996, 207). Yet it might be argued that from a Darwinian viewpoint, homosexuality *is* a kind of malfunction, since it hinders reproductive success. Hence, a therapy/enhancement objectivist could (arguably) characterize altering someone's sexuality from gay to straight as actually therapeutic.

After all, when we judge the health of nonhuman animals, we may look to such things as their fitness to engage successfully within their native habitats in the activities of survival and reproduction, including (in species where it is relevant) the rearing of young (Foot 2001, 33). It might also make perfectly good sense to see physical infertility in humans as an adverse departure from species-typical functioning as well as a common source of anguish and despair. It takes little imagination to see that *bisexuality* might turn out to be adaptive in some circumstances, if only as a sexual outlet in situations where a partner of the opposite sex is unavailable for some reason. A strictly *homosexual* orientation, however, would tend to preclude reproduction and so be nonadaptive.

That is not an inevitable judgment. If we make the assumption that there are genetic potentials for homosexuality, we might still be able to imagine how the relevant DNA sequences could have been selected by evolution. Perhaps they also had the effect of inclining animals with no (or fewer than usual) offspring to care more for the welfare of their kinship group as a whole. In that case, a "gay gene" (a set of DNA variations coding for homosexuality) might actually contribute to the reproductive fitness of homosexual organisms (Murphy 2012, 60–61). But unless such an effect can be substantiated, there is a rationale to categorize a homosexual orientation (or perhaps whatever physical states might underlie it) as an adverse departure from species-typical functioning.

By now, I hope that many of my readers are feeling troubled or getting annoyed. The possible conclusion I've just described strikes me as absurd, not to mention politically dangerous. But where, if it all, have I gone wrong in my reasoning? It appears that when we make judgments about our own species (at least), we cannot seriously judge someone's sexual orientation and other psychological characteristics to be "unhealthy" merely because they tend to hinder reproductive success. Agar (2004, 80–81) therefore is surely correct that attempting to "cure" homosexuality would be setting an irrelevant goal for medical science, as homosexuality is compatible with a high level of well-being. As Foot (2001, 41) observes, there are many forms of human life that do not involve having children, and even when having them is experienced as a good, this human good has complex familial and social aspects with no equivalent in the nonhuman world.

While we care about the well-being of individuals who are childless against their wishes, as a result of circumstances independent of their own personalities and choices, there are limits to this. Accordingly, it would be unreasonable to regard homosexuality as something like a disease or impairment—and the same applies to other sexual orientations or attitudes that do not conduce to reproductive success but instead are compatible with a flourishing life.

All that said, the wrong in trying to "cure" homosexuality would *not* be that such an intervention could, from a bioengineering perspective, be demonstrated to be an enhancement as opposed to a therapy. The problems with attempting such a misguided cure would rather relate to its direct and socially mediated impacts on vulnerable individuals. In the end, the decision of the psychiatric profession in the United States and elsewhere to cease classifying homosexuality as a disease or impairment was not based squarely on objective biological grounds. It was, at least in part, a political one—and none the worse for that.

Age-Related Decline
The most important single difficulty with the bioengineering approach to a therapy/enhancement distinction may be in applying it to age-related decline or illness. LeRoy Walters and Julie Gage Palmer point out that senile dementia, a condition that is experienced by a certain fraction of the aging population, could be seen as falling within the normal (statistical) range of functioning for human beings above a certain age. As they observe, however, we do not hesitate to medicalize it (Walters and Palmer 1997, 121).

Similar observations may be made about many age-related deteriorations of human organs and organ systems. What should we say about the declining powers of, say, the eye during middle age? What about the debility that typically accompanies the last years of a long life? Recall that my expression of (some) support for objectivist accounts of species-normal functioning was based on the fact that they do rely on something in the natural world: the evolutionary "engineering" of organs and organ systems to work in certain specifiable ways. Yet age brings with it the diminution of what might be regarded as our natural gifts or, less rhetorically, a reduction of functional capacity (Kamm 2009, 106–107). While the loss is normal in one sense—it is experienced in every life, or rather by every person who lives long enough to have the experience—it is also true that there is an objectively identifiable departure from the functioning of the body at its peak of efficiency. Think of a well-engineered machine breaking down from long use.

In his book *Better Than Well*, Carl Elliott discusses a suggestion by the biogerontologist David Gems (apparently in conversation). According to Gems, the problem in growing older is "ontological diminution" or "a flattening of the conditions that sustain our existence." Though Elliott (2003, 286) is far from being a prophet of radical life extension, he writes affectingly about the aging process:

As we move into old age, our senses dim, our minds get slower, our sexual desire diminishes, and our bodies lose their physical capacities. Our experience of the world gradually grows dimmer and narrower. As the end of life grows closer, even the future begins to look constricted. The range of possibilities open to us seems to close up, not only because we cannot do as much as before, but because we do not want to. To be relieved of some desires can be a blessing, of course. But a future completely devoid of desire can seem like an especially depressing prospect. To be the kind of person who does not want to do anything whatsoever strikes many people as the worst kind of fate: a kind of volitional castration, lived out in a purgatory of dead routine.

Any concept of Homo sapiens' normal functioning must take into account that the evolutionary process has engineered different kinds of human bodies. There are obvious differences between male and female bodies as well as those between the bodies of children, adolescents, and adults. We develop over time from infancy, gradually attaining our full powers as we grow to adulthood. The bodies of children do not function just like those of growing teenagers, which do not function just like those of adults in their prime.

Given this, Daniels (1986, 21–28, 33, 52, 226) relativizes normal human functioning to a person's stage of life, and the authors of *From Chance to Choice* (who, as noted earlier, include Daniels) assume with no expression of doubt that an intervention to counteract memory loss with age would be an enhancement, albeit one that might be justified (Buchanan et al. 2000, 153–154). Fukuyama and Furger (2006, 66) even suggest that the line between therapy and enhancement is blurred whenever an eighty-five-year-old is offered heart bypass surgery or chemotherapy for cancer, since this could be seen as, in effect, "an unnatural form of life extension."[2] Again, Kamm (2009, 105–106) remarks that most people would consider it a radical enhancement if we succeeded in dramatically lengthening human life spans and thus prevented the process of aging from interfering with the exercise of all the natural gifts we have hitherto enjoyed.

Such responses indicate that at least some people are inclined to think of attempts to retard aging and its effects as paradigmatic cases of enhancement, and not as mere therapy. Surely, though, there is an ambiguity here. Yes, many or most people would probably think intuitively that it goes beyond therapy if an individual is given the dramatic, unprecedented benefit of living for a thousand years. It nonetheless seems extreme and eccentric to claim that something other than therapy is involved when an elderly person is treated for, say, cancer, heart disease, glaucoma, or osteoporosis. If a view such as that of Daniels entails that treating the diseases of old age should not be considered therapeutic, then (we might well think) so much for that view (Harris 2007, 45).

It is even arguable that radical attempts to sustain human capacities and human life for hundreds of years should be regarded as therapy from an objectivist viewpoint. Living for so long is, admittedly, not only statistically unusual; it is unprecedented. We nevertheless can draw an objective line not by using statistics but instead by regarding the human body as, in effect, an intricately functional (naturally evolved) machine. If that machine breaks down as a result of wear over time and we attempt to restore it, that is not enhancing it beyond what it was previously capable of doing. Accordingly, there is a line that actually exists in nature and that we are not crossing. As I've already mentioned, this kind of bioengineering perspective offers support to objectivism—but an objectivism with perhaps surprising implications.

Then again, how surprising are they really? Note that it is difficult to experience the deterioration of capacities through middle and old age as merely a variation on a theme, as a different kind of quasi-engineered functioning. It is not analogous to growing up or the biological differences between the sexes. Though the decline is statistically usual—indeed, inevitable in past and current circumstances—we have good reason not to regard it as an engineering feature. It is hardly startling that most people employ medical and other means to resist the myriad forms of decline that come with age. Why, then, does anyone try to contend that resistance to age-related decline is a form of enhancement rather than a form of therapy?

The answer may lie more in the field of psychology or sociology than philosophy, but there are some possibilities to consider. First, our intuitions may have been shaped by prescientific cultural understandings of human biological functioning. That is, we draw on traditional ways of thinking that are handed down from generation to generation, but that are difficult to reconcile with ideas of the body as an evolved system of organic subsystems that can be studied from a bioengineering perspective. Our intuitive thinking in particular may owe something to pretheoretical ideas that interference with the aging process would be a morally problematic violation of a "proper" course or shape for a human life. Related to this, perhaps, the facts of aging and limits to longevity are so familiar, and so thoroughly influence social organization and everyday thinking, that their complete or substantial abolition would seem like an extremely radical step—one that would immeasurably alter the human condition (and perhaps violate the background conditions to choice discussed in chapter 5). It thus is one thing to accept therapy for cancer or heart disease if it merely delays an elderly person's death. It is another, psychologically, to imagine restoring that person to youthful health and vigor.

Another important fact is that age overtakes most of us at a roughly uniform pace. By contrast, various diseases, impairments, and injuries cause particular disadvantages to particular people at widely diverse times in their lives. This may make remedying them seem more like a matter of fairness among fellow citizens, whereas aging is a common experience that we tend to look on as the default for those who do not die "early." For some thinkers, this might justify placing a low priority on resistance to aging, since the ordinary aging process itself cannot make anybody less of a normal participant and competitor in social activity over an ordinary lifetime (Buchanan et al. 2000, 74).

Yet efforts to combat age-related decline are not simply enhancements or nontherapeutic. They do not give someone greater capacities either beyond or within the normal human range than they previously enjoyed. Rather, they attempt merely to preserve or restore. Even the aim of an outright "cure" for aging would be indefinite maintenance of the efficient functioning of organs, organ systems, and other components of the active human body. The lesson is that this is all multidimensional, involving issues of biology, cultural history, and reasonable social expectations.

Improving on Evolution's Design

Finally, it is worth stressing one more time that the human body and mind were not literally designed; no scheming superintelligence was involved. The body, including the neurological system and everything dependent on it, evolved through a long period of natural selection.[3] Insofar as the process has fine-tuned us for anything, it is for reproductive success in the environment in which our ancestors evolved—to be able to survive long enough in that environment to find one or more mates, and pass down genetic code to the next and succeeding generations.

Wonderful though human bodies may be, and however impressively our organs may perform their work, they are not necessarily ideal for *any* purpose, and certainly not for pursuing our conscious goals and desires in the current environments where we actually find ourselves. Considered from that perspective, we may have many "design flaws," and some of them may be quite gross in scale.[4] Why, then, should we be content with what the blind processes of biological evolution bequeathed us, and why should we not, at least in principle, wish to improve on the result—where the idea of improvement is explicable in terms of efficiency in achieving what we want, as a knife may be improved by sharpening?

Conclusion

Along with the distinction between therapy and enhancement, the existing bioethical literature makes other crucial distinctions, such as that between alterations to somatic cells and those that are made to the germ line—and thus inheritable by later generations of humans (e.g., Buchanan et al. 2000, 106, 230–232). There is also a distinction to be made in the case of genetic interventions between those that are essentially negative, in that they remove unwanted traits—these will generally be related to disease or impairment—and those that may be thought of as positive or adding to desirable traits (Heyd 1992, 169–170; Kitcher 1996, 191). In the context of genetic interventions, this negative/positive distinction is closely related to the therapy/enhancement distinction (Buchanan et al. 2000, 104–110).

Some other distinctions, however, are not so well recognized. One is the distinction between those that are or are not transgenic. There is also a quite-separate distinction between what lies within and what goes beyond the horizon of widespread human desires, although this line doubtless expands as new possibilities are understood, and it may be rather blurred if the desires of human beings vary greatly. Other important distinctions relate to the different technologies that might be used to alter capacities or human psychological dispositions. For example, we can think of genetic, cybernetic, surgical, pharmaceutical, and other kinds of interventions. Still other distinctions include the following:

- Enhancements that involve improved human functioning within (or only marginally beyond) the historical human range as opposed to those involving superhuman levels of functioning (Kamm 2009, 91–92; Wilkinson 2010, 187–191; Savulescu, Sandberg, and Kahane 2011, 8).

- Enhancements that take the individual concerned to a new level of functioning in contrast to interventions (perhaps not best thought of as enhancements) that aim to preserve the individual's experienced capacities in the face of age-related decline.

- Enhancements of the agent who makes the decision versus enhancements of someone else such as the agent's children.

- Enhancements with varying prospects of generating valuable network effects.

- Enhancement technologies with varying prospects for widespread and rapid diffusion.

This is not intended to be an exhaustive list of the required concepts and distinctions. My emphasis throughout this appendix has been on the sorts of problems that arise if we rely overly on a supposed therapy/enhancement boundary when making moral judgments or deliberating about regulatory policy. The problems include the difficulty of fitting the full range of possible interventions into one category or the other (i.e., either therapy or enhancement); the doubtful benefit of applying the distinction to a range of interventions that involve cognitive or dispositional traits; and the distinction's inadequacy for categorizing interventions to resist age-related decline.

We need to work with a richer vocabulary—of words, ideas, and values—if we are to make meaningful contributions to moral debates and policy formulations about emerging technologies. These include the technologies of genetic choice.

Notes

Chapter 1

1. For more detail, see Blackford 2005.

2. For a brief description, see Blackford 2005, 11. On this occasion, I hope to do the question more justice.

3. Originally titled the Prohibition of Human Cloning Act.

4. See especially Somerville 2000, 55–88. Here, Somerville mounts a sustained attack on both reproductive and therapeutic cloning, and calls for their prohibition.

Chapter 2

1. The paragraphs of this subsection draw on a similar, though slightly more detailed, discussion of the harm principle in Blackford 2012. This includes replies to certain objections that will not be analyzed here.

2. For the classic exploration of such cases, where this kind of problem confounds our ordinary thinking about harm and the allocation of blame, see Parfit 1984, 351–379.

Chapter 3

1. For a brief but cogent account of the problems, see Annas, Andrews, and Isasi 2002.

2. For more on this, see chapter 7.

3. We might conclude, along with Stephen Wilkinson (2010, 197–198), that most examples we could imagine involve a mix of competitive and noncompetitive benefits.

Chapter 4

1. This paragraph and the next three closely follow my earlier account, in Blackford 2010, of Strawson's longstanding views. Where the two accounts differ, the current version should be preferred.

2. *Wisconsin v. Yoder et al.*, 406 U.S. 205 (1972). Elsewhere I have argued that this case was wrongly decided; see Blackford 2012, 160–163.

3. For these various formulations, see Habermas 2003, 14, 26, 29, 40–42, 49–51, 60–61, 72, 78–79.

4. I am not confident that Persson and Savulescu (2012, 112–115) have accurately understood Harris's main point in their rebuttal of his views.

Chapter 5

1. This chapter draws on material first published in Blackford 2006b. The argument here has been considerably deepened and extended.

2. Evidently McKibben has spent all his August afternoons in the northern hemisphere, or assumes that his readers have.

3. For a brief look at Feinberg's terminology, see chapter 6.

4. Which of these are really human standards, as opposed to culture-specific ones? For example, Torbjörn Tännsjö (2009) maintains that the values associated with elite sports are no more than a dispensable cultural atavism.

5. For more on the problems for Agar's view of radical enhancement, see Blackford 2011; 2011–2012.

Chapter 6

1. Much of this chapter's content is adapted from material first published in Blackford 2006a. My views, however, have developed considerably, so what I have written here departs substantially from the earlier article.

2. Achondroplasia is a form of dwarfism caused by a genetic mutation that affects bone growth.

3. All citations of *A Theory of Justice* will contain page references to both the original, much-cited, 1971 edition and the revised edition of 1999. Nothing in the argument turns on differences between these editions.

4. I have borrowed the Kazanistan example from Rawls 1999a, 75.

5. See Hart's (1994, 185–200) look at what he calls "the minimum content of the natural law."

6. On the sodomy statutes, see, most notably, *Lawrence v. Texas*, 539 U.S. 558 (2003).

7. For a discussion of this issue (supporting the concept of same-sex marriage), see Blackford 2012, 131–134.

Chapter 7

1. Kymlicka formulates this intuition slightly differently. In case there is any variation in meaning, I use the expression *the Rawlsian intuition* in the way I have stipulated in the text.

2. Or to be more precise, the identities and so forth, of those whose interests they represent.

3. Note, however, that this discussion relates purely to the possible *psychological* effects of enhancement. None of this gainsays the fact that no one is ever an ultimate self-creator, together with whatever moral conclusions this entails.

Chapter 8

1. As it was titled at that time.

2. One might question whether precautions are required in this particular case. It is not at all obvious what dangers are likely to result from the use of embryonic sex selection in the social conditions prevailing in Western nations.

Appendix

1. Note that there is an antivaccination movement, but it relies on pseudoscientific arguments with no influence on mainstream bioethics.

2. I will resist the temptation to state just how offensive I find this form of words, when I think about it in connection with my own elderly relatives.

3. I do not, by this wording, intend to exclude effects from such mechanisms as genetic drift.

4. For a trenchant critique of the idea that the current human genome represents the highest level of human biological possibility, see Buchanan 2011b, 183–193.

References

Ackerman, Bruce A. 1980. *Social Justice in the Liberal State*. New Haven, CT: Yale University Press.

Agar, Nicholas. 1999. Liberal Eugenics. In *Bioethics: An Anthology*, ed. Helga Kuhse and Peter Singer, 171–181. Malden, MA: Blackwell. First published 1998.

Agar, Nicholas. 2002. *Perfectcopy: Unravelling the Human Cloning Debate*. Cambridge, UK: Icon.

Agar, Nicholas. 2004. *Liberal Eugenics: In Defence of Human Enhancement*. Oxford: Blackwell.

Agar, Nicholas. 2010. *Humanity's End: Why We Should Reject Radical Enhancement*. Cambridge, MA: MIT Press.

Allhoff, Fritz. 2005. Germ-Line Genetic Enhancement and Rawlsian Primary Goods. *Kennedy Institute of Ethics Journal* 15:39–56.

Anderson, Elizabeth S. 1999. What Is the Point of Equality? *Ethics* 109:287–337.

Annas, George J. 2005. *American Bioethics: Crossing Human Rights and Health Law Boundaries*. Oxford: Oxford University Press.

Annas, George J., Lori B. Andrews, and Rosario Isasi. 2002. Protecting the Endangered Human: Toward an International Treaty Prohibiting Cloning and Inheritable Alterations. *American Journal of Law and Medicine* 28:151–178.

Bailey, Ronald. 2005. *Liberation Biology: The Scientific and Moral Case for the Biotech Revolution*. Amherst, NY: Prometheus.

Barclay, Linda. 2003. Genetic Engineering and Autonomous Agency. *Journal of Applied Philosophy* 20:223–236.

Baron, Jonathan. 2006. *Against Bioethics*. Cambridge, MA: MIT Press.

Baylis, Françoise, and Jason Scott Robert. 2004. The Inevitability of Genetic Enhancement Technologies. *Bioethics* 18:1–26.

Blackford, Russell. 2003. Who's Afraid of the Brave New World? *Quadrant* 396 (May): 9–15.

Blackford, Russell. 2005. Human Cloning and "Posthuman" Society. *Monash Bioethics Review* 24, no. 1 (January): 10–26.

Blackford, Russell. 2006a. Dr. Frankenstein Meets Lord Devlin: Genetic Engineering and the Principle of Intangible Harm. *Monist* 89:526–547.

Blackford, Russell. 2006b. Sinning against Nature: The Theory of Background Conditions. *Journal of Medical Ethics* 32:629–634.

Blackford, Russell. 2010. Genetically Engineered People: Autonomy and Moral Virtue. *Politics and the Life Sciences* 29 (1): 82–84.

Blackford, Russell. 2011. Review of *Humanity's End* by Nicholas Agar. *Monash Bioethics Review* 29 (3): 8.1–8.7.

Blackford, Russell. 2011–2012. Enhancement Anxiety. *Free Inquiry* 32, no. 1 (December–January): 22–24.

Blackford, Russell. 2012. *Freedom of Religion and the Secular State*. Malden, MA: Wiley-Blackwell.

Bostrom, Nick. 2005. In Defense of Posthuman Dignity. *Bioethics* 19:202–214.

Bratman, Michael E. 2007. *Structures of Agency: Essays*. New York: Oxford University Press.

Brock, Dan W. 2003. Cloning Human Beings: An Assessment of the Ethical Issues Pro and Con. In *Ethical Issues in Modern Medicine*, ed. Bonnie Steinbock, John D. Arras, and Alex John, 631–643. 6th ed. Boston: McGraw-Hill.

Buchanan, Allen. 2011a. *Better Than Human: The Promises and Perils of Enhancing Ourselves*. Oxford: Oxford University Press.

Buchanan, Allen. 2011b. *Beyond Humanity? The Ethics of Biomedical Enhancement*. Oxford: Oxford University Press.

Buchanan, Allen, Dan W. Brock, Norman Daniels, and Daniel Wikler. 2000. *From Chance to Choice: Genetics and Justice*. Cambridge: Cambridge University Press.

Carroll, Lewis. 2009. *Through the Looking-Glass and What Alice Found There*. Westport, Ireland: Evertype. First published 1871.

Charlesworth, Max. 1993. *Bioethics in a Liberal Society*. Cambridge: Cambridge University Press.

Charo, R. Alta. 1999. Cloning: Ethics and Public Policy. *Hofstra Law Review* 27:503–508.

Christman, John. 2009. *The Politics of Persons: Individual Autonomy and Socio-Historical Selves*. Cambridge: Cambridge University Press.

Coady, C.A.J. 2009. Playing God. In *Human Enhancement*, ed. Julian Savulescu and Nick Bostrom, 155–180. Oxford: Oxford University Press.

Daniels, Norman. 1986. *Just Health Care*. Cambridge: Cambridge University Press.

Dawkins, Richard. 2006. *The God Delusion*. London: Bantam.

Devlin, Patrick. 1965. *The Enforcement of Morals*. London: Oxford University Press.

Dworkin, Gerald. 1988. *The Theory and Practice of Autonomy.* Cambridge: Cambridge University Press.

Dworkin, Ronald. 1996. *Freedom's Law: The Moral Reading of the American Constitution.* Oxford: Oxford University Press.

Dworkin, Ronald. 2000. *Sovereign Virtue: The Theory and Practice of Equality.* Cambridge, MA: Harvard University Press.

Ekman, Paul. 2003. *Emotions Revealed: Understanding Faces and Feelings.* London: Weidenfeld and Nicolson.

Elliott, Carl. 2003. *Better Than Well: American Medicine Meets the American Dream.* New York: W. W. Norton.

Farrelly, Colin. 2002. Genetic Intervention and the New Frontiers of Justice. *Dialogue: Canadian Philosophical Review* 41:139–154.

Farrelly, Colin. 2004a. Genes and Equality. *Journal of Medical Ethics* 30:587–592.

Farrelly, Colin. 2004b. The Genetic Difference Principle. *American Journal of Bioethics* 4 (2): 21–28.

Feinberg, Joel. 1970. *Doing and Deserving: Essays in the Theory of Responsibility.* Princeton, NJ: Princeton University Press.

Feinberg, Joel. 1984. *Harm to Others.* New York: Oxford University Press.

Feinberg, Joel. 1985. *Offense to Others.* New York: Oxford University Press.

Feinberg, Joel. 1986. *Harm to Self.* New York: Oxford University Press.

Feinberg, Joel. 1988. *Harmless Wrongdoing.* New York: Oxford University Press.

Feinberg, Joel. 1992. *Freedom and Fulfillment: Philosophical Essays.* Princeton, NJ: Princeton University Press.

Fenton, Elizabeth. 2006. Liberal Eugenics and Human Nature. *Hastings Center Report* 36 (6): 35–42.

Foot, Philippa. 2001. *Natural Goodness.* Oxford: Oxford University Press.

Fox, Dov. 2007. The Illiberality of "Liberal Eugenics." *Ratio* 20:1–25.

Frankfurt, Harry G. 1989. *The Importance of What We Care About: Philosophical Essays.* Cambridge: Cambridge University Press.

Fukuyama, Francis. 2002. *Our Posthuman Future: Consequences of the Biotechnology Revolution.* London: Profile Books.

Fukuyama, Francis, and Franco Furger. 2006. *Beyond Bioethics: A Proposal for Modernizing the Regulation of Human Biotechnologies.* Washington, DC: Paul H. Nitze School of Advanced International Studies.

Gavaghan, Colin. 2007. *Defending the Genetic Supermarket: Laws and Ethics of Selecting the Next Generation.* London: Routledge-Cavendish.

Glannon, Walter. 2001. *Genes and Future People: Philosophical Issues in Human Genetics.* Boulder, CO: Westview.

Glover, Jonathan. 1977. *Causing Death and Saving Lives.* London: Penguin.

Glover, Jonathan. 1984. *What Sort of People Should There Be?* London: Penguin.

Glover, Jonathan. 1998. *I: The Philosophy and Psychology of Personal Identity*. London: Penguin.

Glover, Jonathan. 2006. *Choosing Children: The Ethical Dilemmas of Genetic Intervention*. Oxford: Oxford University Press.

Green, Ronald M. 2007. *Babies by Design: The Ethics of Genetic Choice*. New Haven, CT: Yale University Press.

Gutmann, Amy, and Dennis Thompson. 1996. *Democracy and Disagreement: Why Moral Conflict Cannot Be Avoided in Politics, and What Should Be Done about It*. Cambridge, MA: Harvard University Press.

Habermas, Jürgen. 2003. *The Future of Human Nature*. Cambridge, UK: Polity Press.

Harris, John. 1998. *Clones, Genes, and Immortality: Ethics and the Genetic Revolution*. Oxford: Oxford University Press.

Harris, John. 2004. *On Cloning*. London: Routledge.

Harris, John. 2007. *Enhancing Evolution: The Ethical Case for Making Better People*. Princeton, NJ: Princeton University Press.

Harris, John. 2011. Moral Enhancement and Freedom. *Bioethics* 25:102–111.

Hart, H.L.A. 1963. *Law, Liberty, and Morality*. Oxford: Oxford University Press.

Hart, H.L.A. 1994. *The Concept of Law*. 2nd ed. Oxford: Clarendon.

Haynes, Roslynn D. 1994. *From Faust to Strangelove: Representations of the Scientist in Western Literature*. Baltimore: Johns Hopkins University Press.

Heyd, David. 1992. *Genethics: Moral Issues in the Creation of People*. Berkeley: University of California Press.

Hocutt, Max. 2000. *Grounded Ethics: The Empirical Bases of Normative Judgments*. New Brunswick, NJ: Transaction Publishers.

Holland, Stephen. 2003. *Bioethics: A Philosophical Introduction*. Cambridge, UK: Polity Press.

Hughes, James. 2004. *Citizen Cyborg: Why Democratic Societies Must Respond to the Redesigned Human of the Future*. Cambridge, MA: Westview.

Humanae Vitae. 1968. Encyclical of Pope Paul VI on the regulation of birth.

Hume, David. 1985. *A Treatise of Human Nature*. London: Penguin. First published 1739–1749.

Hurka, Thomas. 1993. *Perfectionism*. New York: Oxford University Press.

Husak, Douglas. 1987. *Philosophy of Criminal Law*. Totowa, NJ: Rowman and Littlefield.

Huxley, Aldous. 1932. *Brave New World*. London: Chatto and Windus.

Juengst, Eric T. 2000. What Does Enhancement Mean? In *Enhancing Human Traits: Ethical and Social Implications*, ed. Erik Parens, 29–40. Washington, DC: Georgetown University Press.

Juengst, Eric T. 2009. What's Taxonomy Got to Do with It? "Species Integrity," Human Rights, and Science Policy. In *Human Enhancement*, ed. Julian Savulescu and Nick Bostrom, 43–58. Oxford: Oxford University Press.

Kahane, Howard. 1995. *Contract Ethics: Evolutionary Biology and the Moral Sentiments*. Lanham, MD: Rowman and Littlefield.

Kamm, Frances. 2009. What Is and Is Not Wrong with Enhancement? In *Human Enhancement*, ed. Julian Savulescu and Nick Bostrom, 91–130. Oxford: Oxford University Press.

Kane, Robert, ed. 2011. *The Oxford Handbook of Free Will*. 2nd ed. Oxford: Oxford University Press.

Kass, Leon R. 2001. Preventing a Brave New World: Why We Should Ban Human Cloning Now. *New Republic*, May 21, 30–39.

Kass, Leon R. 2002. *Life, Liberty, and the Defense of Dignity: The Challenge for Bioethics*. San Francisco: Encounter Books.

Kekes, John. 1997. *Against Liberalism*. Ithaca, NY: Cornell University Press.

Kekes, John. 2003. *The Illusions of Egalitarianism*. Ithaca, NY: Cornell University Press.

Kitcher, Philip. 1996. *The Lives to Come: The Genetic Revolution and Human Possibilities*. London: Penguin.

Kurzweil, Ray. 2005. *The Singularity Is Near: When Humans Transcend Biology*. London: Viking.

Kymlicka, Will. 2002. *Contemporary Political Philosophy: An Introduction*. 2nd ed. Oxford: Oxford University Press.

Law, Stephen. 2007. *The War for Children's Minds*. London: Routledge.

Lindsay, Ronald A. 2005. Enhancements and Justice: Problems in Determining the Requirements of Justice in a Genetically Transformed Society. *Kennedy Institute of Ethics Journal* 15:3–38.

Locke, John. 1983. *A Letter concerning Toleration*. Indianapolis, IN: Hackett Publishing. First published 1689.

Loftis, J. Robert. 2005. Germline Enhancement of Humans and Nonhumans. *Kennedy Institute of Ethics Journal* 15:57–76.

Mackie, J. L. 1977. *Ethics: Inventing Right and Wrong*. London: Penguin.

Malik, Kenan. 2000. *Man, Beast, and Zombie: What Science Can and Cannot Tell Us about Human Nature*. London: Weidenfeld and Nicolson.

Mameli, Matteo. 2007. Reproductive Cloning, Genetic Engineering, and the Autonomy of the Child: The Moral Agent and the Open Future. *Journal of Medical Ethics* 33:87–93.

McKibben, Bill. 2003. *Enough: Genetic Engineering and the End of Human Nature*. London: Bloomsbury.

Mehlman, Maxwell J. 2003. *Wondergenes: Genetic Enhancement and the Future of Society*. Bloomington: Indiana University Press.

Mehlman, Maxwell J. 2005. Genetic Enhancement: Plan Now to Act Later. *Kennedy Institute of Ethics Journal* 15:77–82.

Mehlman, Maxwell J., and Jeffrey R. Botkin. 1998. *Access to the Genome: The Challenge to Equality*. Washington, DC: Georgetown University Press.

Metzler, Ingrid. 2011. Between Church and State: Stem Cells, Embryos, and Citizens in Italian Politics. In *Reframing Rights: Bioconstitutionalism in the Genetic Age*, ed. Sheila Jasanoff, 105–124. Cambridge, MA: MIT Press.

Meyers, Diana Tietjens. 2004. *Being Yourself: Essays on Identity, Action, and Social Life*. Lanham, MD: Rowman and Littlefield.

Mill, John Stuart. 1910. *Utilitarianism, On Liberty, and Considerations of Representative Government*. London: Dent.

Mill, John Stuart. 1974. *On Liberty*. London: Penguin. First published 1859.

Mill, John Stuart. 1998. *Three Essays on Religion*. Amherst, MA: Prometheus. First published 1874.

Murphy, Timothy F. 2012. *Ethics, Sexual Orientation, and Choices about Children*. Cambridge, MA: MIT Press.

Newman, Stuart A. 2005. The Perils of Human Developmental Modification. In *Rights and Liberties in the Biotech Age: Why We Need a Genetic Bill of Rights*, ed. Sheldon Krimsky and Peter Shorett, 203–208. Lanham, MD: Rowman and Littlefield.

Norman, Richard. 1996. Interfering with Nature. *Journal of Applied Philosophy* 13:1–11.

Nozick, Robert. 1974. *Anarchy, State, and Utopia*. New York: Basic Books.

Oliner, Samuel P., and Pearl M. Oliner. 1988. *The Altruistic Personality: Rescuers of Jews in Nazi Europe*. New York: Free Press.

Parens, Erik, ed. 2000a. *Enhancing Human Traits: Ethical and Social Implications*. Washington, DC: Georgetown University Press.

Parens, Erik. 2000b. Is Better Always Good? The Enhancement Project. In *Enhancing Human Traits: Ethical and Social Implications*, ed. Erik Parens, 1–28. Washington, DC: Georgetown University Press.

Parfit, Derek. 1984. *Reasons and Persons*. Oxford: Clarendon.

Pence, Gregory E. 1998. *Who's Afraid of Human Cloning?* Lanham, MD: Rowman and Littlefield.

Pence, Gregory E. 2000. *Re-Creating Medicine: Ethical Issues on the Frontiers of Medicine*. Lanham, MD: Rowman and Littlefield.

Pence, Gregory E. 2004. *Cloning after Dolly: Who's Still Afraid?* Lanham, MD: Rowman and Littlefield.

Pereboom, Derk, ed. 2009. *Free Will*. 2nd ed. Indianapolis, IN: Hackett.

Persson, Ingmar, and Julian Savulescu. 2011. Unfit for the Future? Human Nature, Scientific Progress, and the Need for Moral Enhancement. In *Enhancing Human Capacities*, ed. Julian Savulescu, Ruud ter Meulen, and Guy Kahane, 486–500. Malden, MA: Wiley-Blackwell.

Persson, Ingmar, and Julian Savulescu. 2012. *Unfit for the Future: The Need for Moral Enhancement*. Oxford: Oxford University Press.

Pojman, Louis P. 2000. *Life and Death: Grappling with the Moral Dilemmas of Our Time*. 2nd ed. Belmont, CA: Wadsworth.

President's Council on Bioethics. 2003. *Beyond Therapy: Biotechnology and the Pursuit of Happiness.* New York: Regan Books.

Rawls, John. 1971. *A Theory of Justice.* Cambridge, MA: Harvard University Press.

Rawls, John. 1999a. *The Law of Peoples: With "The Idea of Public Reason Revisited."* Cambridge, MA: Harvard University Press.

Rawls, John. 1999b. *A Theory of Justice.* Rev. ed. Cambridge, MA: Harvard University Press.

Rawls, John. 2001. *Justice as Fairness: A Restatement.* Cambridge, MA: Harvard University Press.

Raz, Joseph. 1986. *The Morality of Freedom.* New York: Oxford University Press.

Resnik, David. 1993. Debunking the Slippery Slope Argument against Human Germ-Line Therapy. *Journal of Medicine and Philosophy* 19:23–40.

Robertson, John A. 1994. *Children of Choice.* Princeton, NJ: Princeton University Press.

Sadurski, Wojciech. 1985. *Giving Desert Its Due: Social Justice and Legal Theory.* Boston: Dordrecht.

Sandberg, Anders, and Julian Savulescu. 2011. The Social and Economic Impacts of Cognitive Enhancement. In *Enhancing Human Capacities,* ed. Julian Savulescu, Ruud ter Meulen, and Guy Kahane, 92–112. Malden, MA: Wiley-Blackwell.

Sandel, Michael J. 2007. *The Case against Perfection: Ethics in the Age of Genetic Engineering.* Cambridge, MA: Belknap Press.

Sasaki, Erika, Hiroshi Suemizu, Akiko Shimada, Kisaburo Hanazawa, Ryo Oiwa, Michiko Kamioka, Ikuo Tomioka, et al. 2009. Generation of Transgenic Non-Human Primates with Germline Transmission. *Nature* 459:523–527.

Savulescu, Julian. 2001. Procreative Beneficence: Why We Should Select the Best Children. *Bioethics* 15:413–426.

Savulescu, Julian, and Nick Bostrom, eds. 2009. *Human Enhancement.* Oxford: Oxford University Press.

Savulescu, Julian, Anders Sandberg, and Guy Kahane. 2011. Well-being and Enhancement. In *Enhancing Human Capacities,* ed. Julian Savulescu, Ruud ter Meulen, and Guy Kahane, 3–18. Malden, MA: Wiley-Blackwell.

Savulescu, Julian, Ruud ter Meulen, and Guy Kahane, eds. 2011. *Enhancing Human Capacities.* Malden, MA: Wiley-Blackwell.

Shelley, Mary. 1992. *Frankenstein, or, the Modern Prometheus.* London: Penguin. First published 1818.

Sher, George. 1987. *Desert.* Princeton, NJ: Princeton University Press.

Sher, George. 1997. *Approximate Justice: Studies in Non-Ideal Theory.* Lanham, MD: Rowman and Littlefield.

Sher, George. 2005. *In Praise of Blame.* Oxford: Oxford University Press.

Silver, Lee M. 1999. *Remaking Eden: Cloning, Genetic Engineering, and the Future of Humankind.* London: Phoenix-Orion.

Singer, Peter. 1993. *Practical Ethics*. 2nd ed. Cambridge: Cambridge University Press.

Singer, Peter. 1999. *A Darwinian Left: Politics, Evolution, and Cooperation*. London: Weidenfeld and Nicolson.

Singer, Peter. 2009. Parental Choice and Human Improvement. In *Human Enhancement*, ed. Julian Savulescu and Nick Bostrom, 277–289. Oxford: Oxford University Press.

Somerville, Margaret. 2000. *The Ethical Canary: Science, Society and the Human Spirit*. New York: Viking.

Somerville, Margaret. 2007. *The Ethical Imagination: Journeys of the Human Spirit*. Melbourne: Melbourne University Press.

Starr, Paul. 2007. *Freedom's Power: The True Force of Liberalism*. New York: Basic Books.

Stephen, James Fitzjames. 1993. *Liberty, Equality, Fraternity*. 2nd ed. Indianapolis, IN: Liberty Classics. First published 1874.

Stock, Gregory. 2002. *Redesigning Humans: Our Inevitable Genetic Future*. Boston: Houghton Mifflin.

Strawson, Galen. 2010. *Freedom and Belief*. Rev. ed. Oxford: Oxford University Press.

Tännsjö, Torbjörn. 2009. Medical Enhancement and the Ethos of Elite Sport. In *Human Enhancement*, ed. Julian Savulescu and Nick Bostrom, 315–326. Oxford: Oxford University Press.

Ten, C. L. 1980. *Mill on Liberty*. Oxford: Clarendon.

Walters, LeRoy, and Julie Gage Palmer. 1997. *The Ethics of Human Gene Therapy*. New York: Oxford University Press.

Wikler, Daniel. 2009. Paternalism in the Age of Cognitive Enhancement: Do Civil Liberties Presuppose Roughly Equal Mental Ability? In *Human Enhancements*, ed. Julian Savulescu and Nick Bostrom 341–355. Oxford: Oxford University Press.

Wilkinson, Stephen. 2010. *Choosing Tomorrow's Children: The Ethics of Selective Reproduction*. Oxford: Oxford University Press.

Wilmut, I., A. E. Schnieke, J. McWhir, A. J. Kind, and K.H.S. Campbell. 1997. Viable Offspring Derived from Fetal and Adult Mammalian Cells. *Nature* 385:810–813.

Wilson, Edward O. 1998. *Consilience: The Unity of Knowledge*. London: Little, Brown.

Index

Basic Bioethics
Arthur Caplan, editor

Books Acquired under the Editorship of Glenn McGee and Arthur Caplan

Peter A. Ubel, *Pricing Life: Why It's Time for Health Care Rationing*

Mark G. Kuczewski and Ronald Polansky, eds., *Bioethics: Ancient Themes in Contemporary Issues*

Suzanne Holland, Karen Lebacqz, and Laurie Zoloth, eds., *The Human Embryonic Stem Cell Debate: Science, Ethics, and Public Policy*

Gita Sen, Asha George, and Piroska Östlin, eds., *Engendering International Health: The Challenge of Equity*

Carolyn McLeod, *Self-Trust and Reproductive Autonomy*

Lenny Moss, *What Genes Can't Do*

Jonathan D. Moreno, ed., *In the Wake of Terror: Medicine and Morality in a Time of Crisis*

Glenn McGee, ed., *Pragmatic Bioethics, 2nd Edition*

Timothy F. Murphy, *Case Studies in Biomedical Research Ethics*

Mark A. Rothstein, ed., *Genetics and Life Insurance: Medical Underwriting and Social Policy*

Kenneth A. Richman, *Ethics and the Metaphysics of Medicine: Reflections on Health and Beneficence*

David Lazer, ed., *DNA and the Criminal Justice System: The Technology of Justice*

Harold W. Baillie and Timothy K. Casey, eds., *Is Human Nature Obsolete? Genetics, Bioengineering, and the Future of the Human Condition*

Robert H. Blank and Janna C. Merrick, eds., *End-of-Life Decision Making: A Cross-National Study*

Norman L. Cantor, *Making Medical Decisions for the Profoundly Mentally Disabled*

Margrit Shildrick and Roxanne Mykitiuk, eds., *Ethics of the Body: Postconventional Challenges*

Alfred I. Tauber, *Patient Autonomy and the Ethics of Responsibility*

David H. Brendel, *Healing Psychiatry: Bridging the Science/Humanism Divide*

Jonathan Baron, *Against Bioethics*

Michael L. Gross, *Bioethics and Armed Conflict: Moral Dilemmas of Medicine and War*

Karen F. Greif and Jon F. Merz, *Current Controversies in the Biological Sciences: Case Studies of Policy Challenges from New Technologies*

Deborah Blizzard, *Looking Within: A Sociocultural Examination of Fetoscopy*

Ronald Cole-Turner, ed., *Design and Destiny: Jewish and Christian Perspectives on Human Germline Modification*

Holly Fernandez Lynch, *Conflicts of Conscience in Health Care: An Institutional Compromise*

Mark A. Bedau and Emily C. Parke, eds., *The Ethics of Protocells: Moral and Social Implications of Creating Life in the Laboratory*

Jonathan D. Moreno and Sam Berger, eds., *Progress in Bioethics: Science, Policy, and Politics*

Eric Racine, *Pragmatic Neuroethics: Improving Understanding and Treatment of the Mind-Brain*

Martha J. Farah, ed., *Neuroethics: An Introduction with Readings*

Jeremy R. Garrett, ed., *The Ethics of Animal Research: Exploring the Controversy*

Books Acquired under the Editorship of Arthur Caplan

Sheila Jasanoff, ed., *Reframing Rights: Bioconstitutionalism in the Genetic Age*

Christine Overall, *Why Have Children? The Ethical Debate*

Yechiel Michael Barilan, *Human Dignity, Human Rights, and Responsibility: The New Language of Global Bioethics and Biolaw*

Tom Koch, *Thieves of Virtue: When Bioethics Stole Medicine*

Timothy F. Murphy, *Ethics, Sexual Orientation, and Choices about Children*

Daniel Callahan, *In Search of the Good: A Life in Bioethics*

Robert Blank, *Intervention in the Brain: Politics, Policy, and Ethics*

Gregory E. Kaebnick and Thomas H. Murray, eds., *Synthetic Biology and Morality: Artificial Life and the Bounds of Nature*

Dominic A. Sisti, Arthur L. Caplan, and Hila Rimon-Greenspan, eds., *Applied Ethics in Mental Health Care: An Interdisciplinary Reader*

Barbara K. Redman, *Research Misconduct Policy in Biomedicine: Beyond the Bad-Apple Approach*

Russell Blackford, *Humanity Enhanced: Genetic Choice and the Challenge for Liberal Democracies*